AN ONLINE DISCUSSION

Visual Culture and Bioscience

ISSUES IN CULTURAL THEORY 12

—

Cultural Programs
of the National Academy
of Sciences
Washington, D.C.

—

Center for Art, Design
and Visual Culture
University of Maryland,
Baltimore County
2008

—

The thoughts and opinions
expressed in this symposium are
those of the authors
and do not necessarily reflect
the positions of the National
Academy of Sciences or
the University of Maryland,
Baltimore County.

MODERATED BY SUZANNE ANKER

With the participation of

- Max Aguilera-Hellweg
- Bergit Arends
- Andrew Carnie
- Oron Catts
- Catherine Chalmers
- Raphael Cuir
- Carl Djerassi
- Florian Dombois
- Troy Duster
- Sabine Flach
- Giovanni Frazzetto
- David Freedberg
- Karl Grimes
- Jens Hauser
- Marvin Heiferman
- Robin Marantz Henig
- Martin Kemp
- Vladimir Mironov
- Leonel Moura
- ORLAN
- Nancy Princenthal
- Ingeborg Reichle
- Miriam Van Rijsingen
- Michael Sappol
- Jill Scott
- Brad Smith
- Andrew Solomon
- Susan Squier
- Eugene Thacker
- Richard Twine
- Catherine Waldby
- Richard Wingate

— Major funding for the symposium was provided by Ralph S. O'Connor and the Marian and Speros Martel Foundation.

—

The National Academy of Sciences (NAS) is a private, nonprofit, self-perpetuating society to which distinguished scholars are elected for their achievements in research, and is dedicated to the furtherance of science and technology and to their use for the general welfare. Upon the authority of the charter granted to it by the Congress in 1863, the NAS has a mandate to advise the federal government on scientific and technical matters. The mission of the office of Cultural Programs of the National Academy of Sciences (CPNAS) is to explore the intersections of art, science, and culture through the presentation of public exhibitions, lectures, and other cultural programs.

CONTENTS

PREFACE

This project affirms our interest at the Center for Art, Design and Visual Culture (CADVC) in expanding the museum's role in examining the social and cultural power of visual images. As part of this effort, *Visual Culture and Bioscience*—a project that attempts to bridge the gap between science and the visual arts—also represents another important aspect of CADVC's mandate: online dialogue. In this regard, this volume and the online conference on which it is based represent the start of a new collaborative effort between CADVC and the Cultural Programs of the National Academy of Sciences (CPNAS) in Washington, D.C.

This book poses some basic questions about the recent intersection between art and science. It does not pretend to be comprehensive, but rather it centers on a range of issues that explore this connection. It will not account for most of the scientific breakthroughs and discoveries that have taken place in recent years; nor will it outline the various political or ethical arguments for or against certain forms of science and their impact on our culture. Yet these, and many other issues and challenges, inform this book's extended dialogue in ways that are once impressionistic, incisive, and artful. This dialogue is much less focused on the outcome of what happens when science and art intersect; instead, it is principally centered on the process and the thinking that lead to the end result.

I am particularly delighted about our new collaboration with the CPNAS. More than two decades ago, I became personally invested in the intersection between art and science. I created artwork that explored this connection, focusing on the notion of science's complicity in constructing historical "lies" and the question of how "truth" is constituted in scientific (as well as artistic) discourse.

Today, artists invested in scientific discourse contribute greatly to a vital, multidisciplinary discussion. In the end, the dialogue in this book poses a number of significant questions: Why are many scientists interested in the arts? Why do artists—over many centuries—continually return to the sciences for content and inspiration? How can we reconcile scientific "truths" based on empiricism and the more impressionistic "truths" of art? To what extent is art influenced by science itself a form of science fiction? To what extent is this art a social or political response to science? What are the differences between the methodologies of the inquiring scientist and the inquiring artist? How do their creative approaches differ?

—

It is my hope that some of the unanswered questions that arise in *Visual Culture and Bioscience* will reemerge in our next collaboration with the CPNAS: an online symposium and publication focused on the issue of "evolution."

David Yager, Distinguished Professor
Wilson H. Elkins Professor
Executive Director,
Center for Art, Design and Visual Culture
Director, Innovation and Design Lab
University of Maryland, Baltimore County

FOREWORD

On January 8, 2008, the National Academy of Sciences and the Institute of Medicine held a press conference announcing the release of *Science, Evolution, and Creationism*,[1] a book that provides a comprehensive picture of the current scientific understanding of evolution. At the conference, the committee chair, Dr. Francisco J. Ayala,[2] stated, "There is now a consensus that biology will be the science of the twenty-first century just as physics was the science of the twentieth century. There are now more scientists working in biology than are working in physics, and the budgets are also larger." This statement underscores the impact that bioscience has had on our lives and is predicted to have on our society in both the immediate and far-reaching future. At the turn of the last century, physics altered life beyond imagination with the discovery of X-rays, the identification of electrons, and the development of Einstein's theory, to name only a few events. Dr. Ayala's statement suggests that the biosciences have brought us to the brink of another exciting era of possibilities.

Shifts in perceptions initiated by advancements in science have always been of interest to artists. Like artists who were confronted by radically new concepts of space and matter during the last century,[3] artists of the twenty-first century have a new muse in bioscience. But does inspiration flow both ways between the arts and sciences? In what way does visual culture contribute to scientific processes? What benefit can be derived from a discussion between artists and scientists? A growing interest in an inquiry of this nature combined with the dynamic impact of the biosciences on our culture prompted the creation of the online gathering "Visual Culture and Bioscience." The dialogue surrounding art, or "visual culture," a term used here for the sake of a more inclusive discussion, provides an opportunity for a multidisciplined expression of ideas, perceptions, and issues that may not fit comfortably within that of typical scientific discourse.

From March 5 to 15, 2007, the Cultural Programs of the National Academy of Sciences joined with the Center for Art, Design and Visual Culture at the University of Maryland, Baltimore County, in creating the virtual meeting place for experts from a range of disciplines. A group of over thirty panelists were invited, forming an assembly of artists, scientists, historians, ethicists, curators, sociologists, and writers. Because of international interest, the symposium was held online. Over 2,500 public participants logged on to read the live discussion, contributing thoughts and reactions. Internet technology augmented the scope of the discussion by broadening the range of panelists and public participants. From points around the world, the panelists logged in from their labs, studios, offices, field work, and even from other conferences. The intent of this exercise was to discuss, explore, and record the current state of dialogue in play between scientists and creative practitioners. The result is printed on the following pages.

— Editing the symposium transcripts for this publication proved to be an interesting challenge. Often, the logic in electronic discourse like this does not flow in an easily readable sequence. At times, several conversations were occurring in parallel and often extended over days. Despite the random nature of communication, the content of the dialogue turned out to be rich and informative. Like an opera, where many voices can be heard at once, the panelists were free to talk concurrently in cyberspace on a variety of thought threads, weaving a net of ideas facilitated by the format. Although it's strangely liberating in a democratic sense to create an environment where everyone can speak at once, unless one's mind is conditioned to the rhythm and sporadic nature of such discourse, the result can seem disorganized and disconcerting. How do we organize the posts such that it is readable, while staying true to the symposium's complex flow of information – a characteristic that makes online conferencing an interesting and valuable alternative to a physical one? Therefore, these transcripts have been reformatted from the original with sensitivity towards flow, consistency, and clarity. It is my hope, however, that we have succeeded in organizing the text such that it is easily readable, while the incredible intensity and spirit of the open discourse is preserved.

— The online symposium was made possible through the generous support of Ralph S. O'Connor and the Marian and Speros Martel Foundation. It was organized by the Cultural Programs of the National Academy of Sciences with support from the Center for Art, Design and Visual Culture at the University of Maryland, Baltimore County. Many individuals contributed to the success of the symposium. Special gratitude is owed to the facilitator, Suzanne Anker, and the panelists who brought their experience, research, and valuable time to these endeavors.

— The conference was organized with the helpful insight of an advisory committee that included Dr. Harvey V. Fineberg, President of the Institute of Medicine; David Yager, Executive Director of the Center for Art, Design and Visual Culture (UMBC); Symmes Gardner, Director of the Center for Art, Design and Visual Culture (UMBC); and Suzanne Anker. Special thanks is given to Alana Quinn for assisting with conference logistics and to William-John Tudor whose skill and technical expertise kept the conference alive in the ether for almost two weeks.

— I would like to express my appreciation for the support given to this conference by Dr. Ralph J. Cicerone, President of the National Academy of Sciences; his spouse, Dr. Carol M. Cicerone; Dr. Harvey V. Fineberg, President of the Institute of Medicine; Dr. Barbara Schaal, Vice President of the National Academy of Sciences; the National Academy of Sciences' council members; and Ken Fulton, Executive Director of the National Academy of Sciences.

JD Talasek
Editor
Director / Cultural Programs of the
National Academy of Sciences

INTRODUCTION

Biofictions and Biofacts: Staking a Claim in the Biocultural Bank

At once archly antagonistic, reeking with envy, at once estranged bedfellows, yet intimate collaborators, at once trepid and at other times trendoid, the domains of art and science continue to be enlocked in a family romance. As with adolescents relishing in their self-abandoned freedom or the fantastical dream of mysterious discovery, each discipline traverses the cultural landscape, determined to penetrate its vault. Peer review panels, funding sources and their lure of major capital are inevitably associated with these shape-shifting enterprises. Add to this dicey mix the ever-increasing archive of technological re-mixes, special effects, and laboratory feats of sheer awe, and we find ourselves in a time where interactivity in all its guises is the central character of the play. But under the rubric of interdisciplinary cross-pollination, what does this intermix bring to the skeptic's investigatory gaze? And at a time when new media criss-cross a uniformed geography of the global, to what means and ends do art and science collaborations justify formal linking?

Eminent art historian Leo Steinberg in his essay "Art and Science: Do They Need to Be Yoked?" makes reference to art and science's on-and-off-and-on-again relationship. He exemplifies the gifts of Leonardo's cultural offerings as a case in point. For Steinberg, "unlike his surpassed scientific work, Leonardo's artistic creation is unrepeatable, like the life of a man."[4] Unicity and rarity become keywords specially woven into the intertwined tapestry of artist and creation. Is it true that neither by measurement nor repetition nor observation alone do works of art attain their status? Or by what happenstance do nonverifiable additions to knowledge production continue to confound our consciousness? And as product-driven methodologies, what do they so ardently demonstrate? Nevertheless, under present parlance, Steinberg's skepticism of the links between science and art may, in fact, be newly up-ended.

Discourses concerning the "in-between-ness" of categories have circulated within the corridors of visual and critical studies for decades. From Rosalind Krauss to Julia Kristeva to Donna Haraway and W.J.T. Mitchell, expanded intellectual frameworks have been useful in analyzing contemporary hybrid practices in the visual arts. Most currently, German philosopher Nicole Karafyllis investigates the resultant forms of hybridization concerning living forms and biotechnological interventions. Coining the term "biofacts" as a neologism combining biology and artifact, she discusses this cojoining as "a hermeneutic concept which allows to ask for the differences between 'nature' and 'technology' in the area of the living."[5] She questions whether the classical distinction between *De anima* and *techne* "still holds true today in light of recent advances in biological and biomedical technologies." For example, to what taxongenomic order does Onco or Rhino mouse belong? As living mouse models, fabricated within the stainless steel and glass laboratory, these sentient creatures come into being by slicing, dicing, "knocking out," or otherwise redistributing their hereditary material. These "biofacts" are the products of a nature/culture transfusion. Although Karafyllis maintains that distinctions between "life" and "technology" still presently hold true, she also continues to think that such distinctions "are much more hidden than before, through the design of living objects in the laboratory."[6]

— The yoking of art and science has produced both novel forms of art and has added an aesthetic dimension to science. Artists are engaging in creative research within scientific laboratories while science institutions employ highly skilled design teams to create visually compelling imagery. However, what significance do these connections, in fact, yield? Is it accurate to speculate that more recently we are finding an ever-growing cast of players within both the sciences and the arts who are embarking on the intersection of this hybrid discourse? Yet a few are yielding unexpected results. Mark Dion's recent exhibition at the Natural History Museum in London, *Systema Metropolis,* curated by Bergit Arends, as part of the museum's contemporary arts program, is a case in point. Dion's investigative techniques involve both scientific and archeological characteristics, yet are singular methods in and of themselves. Uncovering specimens through "archeological" digs, the artist's "laboratory" practice has unearthed several new organisms. One particularly nontraditional insect collecting method consisted of attaching adhesive paper to the roof of his automobile and driving at high speeds down a London promenade. The flypaper was subsequently sent to the lab to be analyzed, employing all the relevant state-of-the-art scientific techniques. And to the amazement of the skeptical scientists at the museum, several new species were verified, hence catalogued, into scientific nomenclature.

— Since the "Visual Culture and Bioscience" symposium, various exhibitions, events, and experimental projects have transpired. Perhaps most strategically relevant is the exhibition at the Museum of Modern Art *Design and The Elastic Mind.* Curated by Paola Antonelli, this elegantly compacted exhibition parces some of the subjects suggested in the symposium. Revolving around science and design, Antonelli cites "elasticity as a by-product of adaptability and acceleration," a dynamic force required to engage an increasingly complex technological world. Revisiting many of the classical themes of this intersection, such as the means by which the invisible can be rendered accessible, form's relation to function, and issues of scale, particularly micro-scale, current technologies have re-opened contemporary discourses on these subjects to other ends. For example, Elio Caccavale's *MY Bio* (2005) is a collection of toys that introduces children to the ways in which emergent biotechnologies can interface with the future of life. Like the symposium's foray into toys as chimeras and viruses, Caccavale creates a series of human dolls that undergo transplant surgery or cows that produce pharmaceutical drugs. Joris Laarman's *Bone Chair* (2006) employs 3-D optimization software to create a chair that, in this case, is based on the ways in which bones actually grow, thus emphasizing nature's structural competency.[7]

At present, streams of unique projects, processes, and collaborations underlie exploratory and investigatory models for innovation inextricably bounded by intersections between art, science, and technology. Possibility abounds in the pursuit of inquiry which operates in degrees encompassing complexity and ambiguity. Entities that are clearly neither science nor art exclusively but an intricate mix of aspects or divisible ratios between the two are beyond the glitter of neophillia. However, the epistemological underpinnings of knowing and being in the world within these practices are pointing the way towards an arena bounded by the diaphanous. Scientists refer to their spectacular color images as art, and the visual artist's conceptual free zone has added tangible facts to the scientific community. Even though all disciplines possess cultural dimensions that are embedded in society's DNA, it is within the matrix of intersecting epistemologies that other strands of knowledge are being produced. Whether by proximity, serendipity, chance, or egotistical tenaciousness, unexpected convergences are taking form. However, "art-sci" collaborations in and of themselves do not necessarily produce significant work no matter how the coordinates are spinned. While many projects produced under this guise could be considered naïve or without significant merit, there are always a few that break the template.

What is "sci-art"? Or shall I say "art-sci"? As a contraction in its contemporary usage, what is its origin? One place to start is to examine the projects in this area funded by the Wellcome Trust in the UK. Currently under the direction of Dr. Ken Arnold, Head of Public Programs, the Trust's mission is manifested by engaging the public understanding and appreciation of science "through the collections and facilities within the Wellcome Building," completed in 2006. The building also is the site for the "Wellcome Library, two permanent galleries, a temporary exhibition space, as well as auditorium and 'forum' spaces, which accommodate scientific debates, seminars, and drama productions." As a pioneering foundation serving both the scientific and artistic communities, the Wellcome Trust program serves as an initiator of collaborative projects between artists, scientists, and institutions aimed towards innovative ideas in this area.[8]

— Alternatively, Sian Ede, as Arts Director of The Calouste Gulbenkian Foundation in London, funds projects in this mix. Reversing the conjunction of contracted words from "sci-art" to "art-sci," the positions of firstness are reordered. In this sense, it is art that assumes primary role. Sian Ede explains the role of the foundation, which has been funding projects for over ten years. "While we are primarily interested in supporting artists to develop their thinking and practice in order to make new work," she states, "we have also sought to encourage scientists to understand the unusual and imaginative response to the world that artists take, with a view to giving greater credibility to other ways of thinking intelligently (visually, dramatically, kinetically), beyond the cerebral."

— A more recent addition to this junction is David Edwards's *ARTSCIENCE: Creativity in the Post-Google Generation,*[9] a text reinforcing and expanding the mission of the fusion between art and science. As a guest on NPR's *Weekend Edition* segment on Arts and Culture,[10] David Edwards talks about *Le Laboratoire*, a hybrid art-science space he established in Paris in 2007. He envisions this experimental space as a way to bring into the public domain what he calls "innovative intelligence." For Edwards, a middle ground that is "at once aesthetic and scientific—intuitive and deductive, sensual and analytical —is the goal of his concept."[11] Recently at le Laboratoire, a collaboration between Michelin chef Thierry Marx and physicist Jerome Bibette explores the state of matter known as a colloid. Coined by Scottish chemist Thomas Graham in the nineteenth century, a colloid is defined as a type of physical property in which "nondiffusable particles are suspended in a surrounding medium of a different substance." Neither a mixture nor a conglomerate, this state of matter, as a type of emulsion, produces particular textures in food and has other unique properties. "Gastronomie moléculaires" is a foray into the science of microparticles, which, when bursting in one's mouth, reveals the full flavors of haute cuisine.

— Another example of the growing platform within this arena is the exhibition *sk-interfaces: Exploding Borders —Creating Membranes in Art, Technology, and Society*, curated by Jens Hauser for FACT (Foundation for Art and Creative Technologies) in Liverpool, UK. The works in the exhibition "propose a 'skinless society,'" in which skin becomes a metaphorical membrane of osmosis between inside and outside, or a territory with no fixed limits. Bringing into view the work of an international array of artists, this exhibition and its theoretical bent can be virtually revisited by paging through its outstanding catalogue. The catalogue is itself encased within an interactive skin, a cadmium orange, padded, vinyl-like cover that is receptive to touch. Indices of fingermarks and handprints become visible as they collide with the sk-interface surface.[12]

— BRAINWAVE, a New York City-wide art, lecture, and performance series, from January to June of 2008, brought together the Rubin Museum of Art, Exit Art, Science and the Arts at the Graduate Center of the City University of New York, the Philoctetes Center for the Multidisciplinary Study of the Imagination, the School of Visual Arts, and the American Museum of Natural History to showcase the cultural dimensions of the neurosciences. Experts in differing fields ranging from visual artists and scientists to musicians, monks, and magicians engaged in unusual pairings. Tim McHenry of the Rubin Museum of Art sees this collaboration as a platform stimulating "creative synapses firing as cross-disciplinary insights."

— Exit Art, a NYC nonprofit institution, presented *Common Senses,* an exhibition of works of visual art consisting of painting, sculpture, video installation, and recorded fMRIs. From robotic behavior to cellular automata to abandoned photographs of psychiatric patients, the exhibition formed a backdrop for neuroscientist Joseph Le Doux and his band, The Amygdaloids, to perform a rock concert. The band's name is a pun on the word amygdala, a section of the brain whose anatomical structure is connected to behavior that is largely emotionally driven. *Common Senses* is the latest foray of Exit Art into a series of exhibitions, Unknown Territories, which explore the visual arts in relation to science and technology.[13]

— Many new texts have appeared and numbers of symposia continue their calls for papers. From Barbara Maria Stafford's *Echo Objects: The Cognitive Work of Images* (M.I.T. Press, 2007) to Oliver Grau's *New Media Art Histories* (M.I.T. Press, 2007) to Eduardo Kac's collaboration with Avital Ronell, *Life Extreme* (Dis Voir, 2007), an expanded literature on this subject is greeting the reader from myriad directions. And as a further update, artist Steve Kurtz of the Critical Art Ensemble has been exonerated of federal charges against him for the illegal transport of bio-substances via the United States postal service. And at the University of Western Australia,

Perth, a graduate degree in the Biological Arts is being offered. Open to applicants with an undergraduate degree in the sciences or the arts, this program aims to focus on "creative bio-research."

—

For Leo Steinberg, the yoking of art and science remains a skeptical pairing. However, in the accelerating age of pixels, bits, and bytes, images are both computational output and aesthetic barometers. As the partaking of dual identities, images, installations, and laboratory life forges ahead into concepts concerning liminality, ambiguity, and new representational spaces, the epistemological underpinnings of the ways in which visual culture and the bio-sciences form "poly-disciplines" are questions requiring innovative, if not radical perspectives.

Suzanne Anker
Moderator
Chair, Fine Arts Department
School of Visual Arts, New York City

W E L C O M E

JD TALASEK

— On behalf of the Office of Cultural Programs of the National Academy of Sciences and the Center for Art, Design and Visual Culture at the University of Maryland, Baltimore County, I would like to welcome everyone to the online symposium on "Visual Culture and Bioscience."

— Many thanks to Suzanne Anker for her work in planning and facilitating the symposium.

— We are looking forward to a dynamic and fascinating discussion.

SUZANNE ANKER

— Greetings out there and welcome to our symposium, "Visual Culture and Bioscience," sponsored by the National Academy of Sciences and the University of Maryland, Baltimore County. Although we are located in the United States, our discussion inhabits the "nether zone" of everywhere and nowhere, with literally an absent *terra firma* underfoot. Let us assemble wherever we may be and begin our online dialogue. Our international discussion commences on March 5 and concludes on March 13, 2007. Intended as an update on the intersections of distinct, yet overlapping, disciplines in art and the biological sciences, we welcome your thoughts, fleeting or otherwise, as segues into understanding more fully the nature of experimental representational systems and their influence on the social order.

— Over the course of the symposium, we will be addressing three interrelated, broad-based topics but leave open the possibilities to reverse courses as necessary, much the way of any trafficking endeavor.

— Our schedule is as follows:
March 5-8, 2007: Imaging in Art and Science
March 9-10, 2007: Artists in the Lab
March 11-13, 2007: Social and Cultural Implications of Visualizing the Biosciences

MARCH 5–8

SESSION 1

IMAGING IN ART AND SCIENCE

SUZANNE ANKER

— The ubiquitous employment of digital technologies within the practices of research science and medicine, architecture and design, filmmaking and video production, as well as the visual and performing arts, has set ajar a multiplex of communication networks that crisscross traditional boundaries. In doing so, malleable coordinates of perception (in time and space) create vast arrays of alternating conceptions of how to structure the problematics of the twenty-first century's arts in relation to the innovations brought forth by science and technology. At a time when the biosciences may be considered to be experiencing a "golden age," the arts, on the other hand, struggle not with public consumption, but with a more profound challenge to its intrinsic identity and history. What role do the visual arts assume in contemporary discourses of "knowledge production"? What internal striations are evident in the functions of art as entertainment, commodity, and critical practice? How do these aspects intersect or correlate with the contemporary biosciences?

— Twentieth-century art and science shared many defining characteristics: abstraction, fragmentation, and reductionism, to select a few. However, in the twenty-first century, one may inquire, what migratory attractions between these disciplines are currently present? One aspect of art's relationship to science is evident within modes, styles, and devices of visual representation. Digital technologies, part and parcel of all Western-type knowledge-producing institutions, enhance the connective tissues between the studio, the laboratory, the scholar's office, and the writer's den. To employ a linguistic metaphor, these technologies have become, ipso facto, the linqua franca of our time. In addition, intersecting domains of inquiry appear to confront family resemblances with regard to the questions they consider: beauty, instrumentalized vision, data manipulation, et al. Whether conscious or not, these crossovers in visual application, narrative interpretation, and symbolic models of the real continue to exude and reframe the philosophical implications of perception and cognition, authenticity and artifice, in both science and art.

— Pictures, maps, diagrams, drawings, notations, scores, and the like become for their moment a mirror of thought. Whether to record or document, deceive or aggrandize, sell or console, they are unmatched agents of communication. Like the mind itself, sometimes these agents of signification embrace layers of alternating meanings, ambiguously disguised hence subliminally hijacking our nervous system. Or, on a more preeminent day, when media spin is not on parenthetical overload, hints of received clarity or even sudden, subtle epiphanies may muster our reserves of potential energies to engage and consider perplexing questions of the day.

— These are compelling ideas, and I cast them in your direction. Let us see if we can parse the complexities, treacheries, and clichés so overtly familiar to this subject. I am aware that this shoot-from-the-hip approach in real time may be a bit intimidating. Perhaps a discussion of this kind among experts, peers, and practitioners, makes us feel like we inhabit a primal childhood dream: a dream of being seen vulnerable and naked splayed out on a beach with our private parts in clear view to the external world. As we move forward with this symposium, I wish to assure you that your thoughts on this subject are a necessary next step in further elucidating the intricacies of our cognitive and emotional worlds. Therefore, like Carl Sagan, but in earthly form, I ask, is anyone out there?

— With these concerns stated, I present the following questions to our panelists:

1. What role do picturing practices play in your discipline of "knowledge production"?
2. How have your perceptions and attitudes of mind been challenged by current dialogues within the art-sci arenas?
3. What role have new imaging technologies played in your conceptualizations of visual modeling or artistic application?

— Your answers to these questions and issues can be addressed within the realms of your personal experience, empirical evidence, or conceptual understanding.

[public blog]

SUZANNE ANKER

— Dear Bloggers, Data Miners, and Citizens of the Public Domain,

— On behalf of our subjects of interest, we request the pleasure of your company to partake in our discussion, Visual Culture and Bioscience.

— Quibbles, solutions, interrogations, quips, disputes, suspicions, and acute responses may be employed to quicken, strengthen, penetrate, or even activate this discussion. Understated, coy, and restrained opinions unassumingly may add a measured response as well. We seek to clarify rather than pacify or needlessly alarm. We seek to explore and engage elusive issues with regard to visualizations, representations, picturing, and processing in the parlance of image communication and interpretation. Hyperbolic and party-line responders need not apply. Please join us for a lively discussion.

FLORIAN DOMBOIS

— I am an artist with a background in geophysics, investigating still mainly tectonic phenomena and earthquakes, but in artistic forms. I try to depict and formulate research results in different media that usually cannot be named in scientific terms. I don't want to beautify scientific results, and I don't want to borrow dignity or value from scientific research. That kind of illustration I am not interested in.

— In my work, sound plays a main role. Using audification of seismic registrations and other movements, I listen to the earth and its activity and develop from here most of my results. And on a second layer of understanding, I intentionally change my media and forms of depiction to watch how my research topic is changed/affected by the Eigensinn of the medium itself.[15] So far, I have worked with sound and different forms of installations, CD, DVD, concert, virtual reality, and interactive art. I have also published a patent, a picture book, wrote different text genres, gave talks, etc. I am convinced that changing the form is affecting the content.[16]

— I have been pushing the art-sci dialogue since about 1990. I am very happy that the dialogue is now much more accepted than fifteen years ago. But nevertheless, we need to raise the quality of that exchange. It is not enough to bring artists into labs and scientists into ateliers. We need to take the claim seriously: that art also is a form of knowledge production, and draw conclusions from that. There is not enough time here at the symposium to explain the whole topic, and my English is too poor for that. But I am concerned about an "art as research" program that would allow artists to apply for research funding here in Switzerland. Artistic research has a lot of open questions that need to be discussed. Therefore, it is important, I think, to run it as a major movement with many players.

— I have worked for a while at GMD / Fraunhofer Institute, a research institute in Germany (which had the first CAVE[17] in Europe and also invented different display technologies). There I developed (1999-2004) a few virtual reality installations using real-time rendering of sound and images, 3-D images and 3-D sound, and interactive technologies, etc. I think that visualization and sonification are good starting points to convince scientists how much other medias of publication beyond scientific papers in journals can affect their research developments. [18]

CARL DJERASSI

— As a preamble to my comments, I need to point out that I am addressing the three questions from the perspective of a scientist who has turned into a playwright. Therefore my "artistic application" is playwriting, which probably was not even envisaged by the organizers of this symposium.

— I have now written eight plays—six for the commercial theater and two for pedagogic (classroom) venues—all of which can be downloaded together with performance schedules, reviews, etc.[19]

— All of these plays have a didactic component in terms of "information transmission," which is frowned upon by many theater scholars as well as practitioners, many of them inherently science-phobic. But since I have chosen to ignore that opposition, visual representation of "practices" are important in my plays. I shall cite just one example:

— In my play *An Immaculate Misconception*,[20] which by now has been translated into eleven languages, the main theme is the practical use of new assisted reproductive IVF [in vitro fertilization] techniques and their ethical implications—a theme that is also implied in the play's subtitle, *Sex in an Age of Mechanical Reproduction*, with its obvious analogy to Walter Benjamin's *The Work of Art in an Age of Mechanical Reproduction*. In my play, I discuss ICSI [intracytoplasmic sperm injection]—the direct injection of a single sperm into an egg under the microscope. In real life, the microscope is usually connected to a video monitor, and this is also done in my play, where the scientist on stage, who is conducting the supposedly first ICSI injection into a human egg, debates this with a physician while the latter observes the procedure on a huge projection wall on the stage facing the audience. In addition, e-mail exchanges and blogs between the scientist and the putative sperm donor are also projected in real life as if they were just typed out.

— Again, while the art-sci definition in the above question is probably not meant in the context of theater, I have addressed that very specifically in terms of "what can theater do for science" vs. "what can science do for the theater." Since I have just written a longish article on that topic, "When is Science on Stage Really Science?" in the January 2007 issue of *American Theatre*, I refer interested persons to the following web site.[21]

[public blog]

SEMEIOTICA

— Immaculate Misconception! What a great title! I would love to see the play; it sounds very interesting.

— This series of comments reminds me of the similarities between scientific narrative as it is presented in, for example, journal articles and the dramatic narrative evident in theater and film. The common narrative arc for science (introduction, methods, results, discussion, conclusion) depends very much on cause and effect and follows closely the style of narration in film (exposition, some change in knowledge, a goal-oriented plot, investigation, and, finally, the climax).

— I'm curious about the didactic quality often associated with information transmission and its role in pedagogy. Alfred Hitchcock remarked that "suspense is the most powerful means of holding onto the viewer's attention.... It is indispensable that the public be made perfectly aware of all the facts involved.... [The] conditioning of the viewer is essential to the buildup of suspense."

— Suspense is vital to narration in theater and film and is implicitly embedded in scientific communication both within the discourse of science and between researchers and the "public" audience.

— One might argue that the usual style is a broad kind of suspense that keeps the audience in the dark about what will happen next and creates uncertainty. This kind focuses only on the protagonists, so that when anything significant happens, it is a surprise to the viewer. As such, their responses can vary more widely, depending their prior knowledge, which may or may not prepare them.

— A second kind of suspense keeps the audience attentive through the use of deadlines and frequent shifts in perspective from the protagonist to other "actors," human or nonhuman.

— I'd be curious to hear what theater folks and others who deal with stories and plot structures have to say about the use of these tactics to shape and moderate scientific narration.

SUSAN SQUIER

— Carl Djerassi writes that his use of picturing practices has the didactic component of "information transmission." I view the function of picturing practices somewhat differently. I understand images to be conveying in condensed form the tangle of human relations, practices, wishes, desires, and fears that pose social, political, and ethical challenges to scientific practice. Such condensed images, or as Ludwig Fleck[22] terms them, ideograms, serve a particularly important role as sites where alien and archaic thought styles are preserved even into our modern era. As Fleck points out, "To the unsophisticated research worker limited by his own thought style, any alien thought style appears like a free flight of fancy, because he can see only that which is active and almost arbitrary about it." Yet whether they are ancient or modern, ideograms are essential to science, as part of the process of knowledge production and knowledge dissemination: "There is no visual perception except by ideovision and there is no other kind of illustration than ideograms."[23]

— To turn to a specific and curiously overdetermined example, in a 1995 essay I explore a photographic image, by Jock

McDonald, of a pregnant Carl Djerassi *[fig. 1]*. I argue that this photograph, published in *Stanford Magazine*, not only provides us with a witty visual play on Djerassis's role as father of the birth control pill. It speaks to a powerful fantasy fueling the technique of assisted reproduction: the wish to usurp female procreative power. This fantasy, Evelyn Fox Keller[24] has shown us, is intimately linked to the rise of modern science. It has been explored in literature as long ago as Mary Shelley's *Frankenstein* and as recently as Angela Carter's blistering *The Passion of New Eve*. And as I have demonstrated in *Babies in Bottles*,[25] it is a fantasy frequently invoked by scientists, fiction writers, and the popular press in the years during which the building blocks of assisted reproduction were being developed: tissue culture, embryo culture, laparoscopy, and in vitro fertilization.

— Lest it seem that literature is the only place such fantasies are explored, I also point out in my article that McDonald's image was preceded by a public policy debate on the same question. In 1989, the Australian National Bioethics Consultative Committee issued a discussion paper, "Developments in the Health Field with Bioethical Implications." In a section on "Transsexuals and Abdominal Pregnancy," the paper considered whether a case could be made for male pregnancy as a bioethical right. The argument for male pregnancy as a right for transsexual men envisioned a time when abdominal pregnancy was possible for biological men, including those who had undergone sex reassignment surgery. The right to a male abdominal pregnancy could then be upheld, the argument hypothesized, as part of one's right to reproductive autonomy. The counter argument held pregnancy to be a collective choice bearing significant social responsibilities:

It may be that the only possible approach to the potential demand for abdominal pregnancy is to focus not on the desires of individuals but on the social and political questions posed by resource allocation and the ethical questions posed by the appeal to a technological imperative.[26]

— As my use of McDonald's photograph in my article demonstrates, I understand the image of the pregnant Carl Djerassi as more than a humorous, topical riff on scientific reality. To my way of thinking, it does indeed convey information, but not precisely in the mimetic sense that Djerassi's post suggests. Rather, the information conveyed is closer to that contained in Fleck's ideograms: the rich, complex mixture of socially, politically, and personally freighted tensions and desires that gave rise to, and now issue from, the scientific practices we know as assisted reproduction.

CARL DJERASSI

— Squier raises an interesting and valid point (although it's not an "either or" but rather "as well as" information transmittal). It is ironic that I have forgotten the McDonald photo of me as a pregnant man, which at the time was done for its "shock" value rather than the more sophisticated interpretation mentioned by Squier. Interestingly, when the German translation of my autobiography *The Pill, Pygmy Chimps, and Degas' Horse*, appeared, it bore the title *Die Mutter der Pille*[27] and carried the McDonald photo of me on the spine. In this case, the meaning is considerably more complicated in the Squirean sense, since I make the point in that autobiography that, in my opinion, the chemist is always the mother of a synthetic drug and the biologist the father (with the clinician the midwife), since nothing can be done by the biologist until the chemist has first conceived (pun intended) of the chemical structure of the drug and has then synthesized it. In that sense, the chemical substance is the egg and the biological experiments the sperm, with the key biological experiment being the sperm that achieves fertilization of the egg.

— And returning once more to the example I gave earlier, where, in my play *An Immaculate Misconception*, a filmed ICSI (single sperm injection) of an egg is projected, its use in the Squirean sense was extended in the French production of my play in Geneva, where the ICSI injection itself was projected onto an egg-shaped surface, and this in turn was constructed in such a fashion that the love scene between the (unwitting) sperm donor and the female reproductive biologist was performed within that egg. I am attaching three pictures from the production to illustrate that point *[figs. 2, 3, 4]*.

SUZANNE ANKER

— To pick up on Carl Djerassi's descriptive introduction of himself as "scientist turned playwright" and in keeping with previous questions with regard to comparisons between linguistic and iconic modes of communication, I relay to you some of Nelson Goodman's parenthetical descriptions of the structural differences inherent in various art forms.

— In *Languages of Art: An Approach to a Theory of Symbols*, published in 1976,[28] he differentiates between works of art as being autographic or non-autographic. To those works that are categorized as non-autographic, he assigns the term "allographic." In this differentiation, which, in fact, may or may not be extrapolated to include laboratory science, an autographic work is a fait accompli, executed by an artist with no further elaboration or revision. Its solitary stance as an object in the world is its unique stature. For the allographic arts, such as the musical and dramatic arts, there is a two-step externalization process at hand. The scripts, scores, or notations continue over time as fixed elements but simultaneously have the capacity to incorporate manifold varieties of future productions and interpretations. Hence in allographic works, alternative "readings" become part of the aesthetic output.

— Carl Djerassi also points out that many "theater scholars as well as practitioners are inherently science-phobic." I must agree with this interpretation, particularly as it relates to the visual arts. What baffles me, however, is why scientific metaphors are marginalized in art world parlance, while often trite re-runs of "received ideas" are applauded. Even images that embed questionable social narratives are applauded for their ironic stance. Why is banality ironic?

CATHERINE WALDBY

— Hi ya,

— I often see a relationship between digital imaging and its endless mutability and tissue engineering, rDNA [reconbinant deoxyribonucleic] technology, and the mutability of organisms. Sometimes digital animation, manga, and sci-fi images appear to act like testing grounds or articulations of biology's new aspirations to endlessly reorder morphology, to tinker with embryogenesis, to both cause and prevent mutation, etc.

— It is a relationship well exemplified by Patricia Piccinini's work[29]—her very obviously synthesized images of cute but very disturbing petlike creatures, who are themselves both synthesized and domesticated somehow – the viewer is left to imagine their provenance *[fig. 5]*.

SUZANNE ANKER

— Catherine,

— Several issues come to mind with regard to the facile mutability of digital images and their counterparts in 3-D computer-generated objects. Certainly Patricia Piccinini's work *We Are All Family*, exhibited at the Venice Biennale in 2003, is an excellent example of how visual images can arouse in the viewer dramatic somatic responses.[30] Can you offer our panelists some additional examples of "triggering" images within the domains of animation, manga, and sci-fi? Perhaps a link or picture attachment would do?

RAPHAEL CUIR

— Hi everyone,

— To follow up on Catherine's example and Suzanne's demand for more, Cronenberg's *eXistenZ* (1999) came to mind. In this film, the "Pod" (a new kind of PlayStation) is directly plugged into the body, which allows players of virtual reality games to enter virtual spaces within the game. The design of the pod is very much related to Piccinini's pets. In the movie, Allegra Geller (Jennifer Jason Leigh), a star game designer, interacts with the pod as if it were a pet: when it is damaged, she expresses feelings—angst—as she would for her favorite cat or so. Humans' relationship with the pod is very complex, as it is connected through a cord which really looks like an umbilical cord. It is related to a pre-Oedipian connection with the mother. The plug in the body is extremely sexualized, especially as Allegra Geller lubes the brand new orifice plug in the body of her fellow partner. The biotechnology image here suggests a regressive experience into a virtual world where matrix seems to have a literal as well as a metaphorical meaning.

MICHAEL SAPPOL

— I'm a historian. I work on the history of anatomical illustration, medical iconography, and representation. I'm also an exhibition curator. A few sketchy thoughts: new imaging technologies (MRI [magnetic resonance imaging], sonography, CT [computed tomography] scan, PET [position emission tomography] scan, etc.)—like anatomical engraving, watercolor, pencil sketching, chromolithography, photomicroscopy, chronographometry—have become a part of our visual environment, a vocabulary deployed by fine artists and designers, in gallery, book, and museum settings, as well as in commercial film, video, print, computer graphics, and other media. Such images become pervasive. There is a cumulative naturalization effect: the digitized, manipulated image, paradoxically, signifies the natural world and becomes the accepted order of things. Even historical productions that might provide a counter-example, such as the pre-computer-age imaging technologies listed above, become subsumed by digitalization, insofar as most people only know such productions through computer, television, or digitized printed pages and projections.

ANDREW CARNIE

— I would like to add a further couple of points on imaging technologies.

— I have heard a photographic assistant say, tongue in cheek, of her boss, a scientist she worked for in a lab, "his current favorite filter is the embossing filter; I wonder what it will be next month?" The sense was that the scientist was slowly learning Photoshop, and the next round of images he might produce might be enhanced by the next "discovery" he made in the Photoshop filter palette.

— Many books have amazing color images from electron microscopes. These images are made from information drawn from a complex process. They are digital and without color. Color is applied to it via filters and histograms, often in Photoshop. This false color is used to enhance a "factor" in the photograph that the scientist wants to exemplify.

— Further, I have found that some images in science imaging competitions can be composites of several images put together, with no disclosure of this fact. They are beautiful, but not truthful.

— I always think there should be better descriptions of science photographs in terms of technical production.

— Another point might be that "images" are often not what science or the scientist is looking for. They act merely as guidance and evidence for what is happening. Pictures and images are not the final product. It is an understanding that matters for them. In scientific papers, they are often secondary in significance. Richard Wingate's QuickTime movies of neurons growing in the chick brain only give clues to the process. It is the process he is interested in—the switching on and off of genes that make the proteins that guide the neurons. They are spectacular sequences, and I use them as an artist because they give a sense of the changing brain. The fact that these cascading flows of neurons all move into their brain "location" in the chick brain in some twenty-two days and that, on hatching, the chick can see and walk almost immediately is incredible.

BRAD SMITH

— Greetings fellow panelists and visitors,

— In terms of my background, I move back and forth between creative visual work and science. I work as an artist, as a researcher in radiology, as a faculty member in art and design, and as a medical illustrator. Recently, it has not been clear to me which role I am playing at any particular moment.

— I used magnetic resonance microscopy in the 1990s to detect and document cardiovascular development in mammalian embryos. This was a time when molecular and developmental biologists were beginning to manipulate many genes thought to be important to development of the embryo. However, these same researchers were not trained or equipped to recognize and interpret the results of their genetic manipulations. They often were not able to distinguish a normal from abnormal embryo looking at it directly or observing it under a microscope. It was in this environment that I developed MRI methods to record the normal and abnormal development of genetically manipulated embryos, with the aim of understanding if the gene in question was essential to normal development. This imaging technology was used to look for very specific morphological evidence of changes from normal development. Optical histology had been used historically, but the single slice images of histology were not adequate to understand complex three-dimensional structures that were changing during development.

— MRI of the embryos allowed me to produce virtual 3-D (and 4-D, with time as the fourth dimension) depictions of developmental processes. I began to use animations of rotations, fly-throughs, and dissolves of the embryos to depict the results of my colleagues' genetic manipulations. Thus, imaging was used to investigate, to document, and to communicate. It certainly had other social consequences as described above and below. Interestingly, this imaging process also posed new biological questions that would not have been considered otherwise, and thus the imaging led to new questions and new experiments. The imaging became part of an iterative process, similar to the iterative process I experience in my creative work, where viewing a creative artifact will suggest unanticipated questions and seek new interpretations.

— It is also curious that imaging mouse embryos exerted a social or political influence sufficient to dislodge a considerable amount of money from a large bureaucracy. The National Institutes of Child Health and Human Development requested the production of an MRI atlas of human embryos because of the images they had seen of mouse embryos.[31]

— While the aim of the human embryo imaging project was to document human embryos for educational, research, and clinical use, the more interesting consequence was a flood of e-mail from the public, with very private and poignant questions about human reproduction, development, and the meaning of being human. The quotations included with this message are responses to seeing MRI depictions of embryos resulting from this funded research. They suggest an emotional influence of imagery that was intended to be clinical.

— The following are e-mail comments to a website showing MRI depictions of embryos. The genesis of the website is explained later in this response:

— "I am now having second thoughts about having an abortion, but I know that I would not be able to give my child a good life."

— "I would like to know if an embryo is also regarded as a living human?"

— "...at what stage does it get the human soul?"

— "I was wondering how I could find out about selling my embryos in the state of Georgia."

— "...He said that I should get an abortion because it is just an embryo right now. Could you tell me exactly what an embryo is?"

— "I just had a frozen embryo transfer done on Tuesday, and we were wondering what our chances of conceiving are."

— I have subsequently revisited these image data as an "artist," addressing ways that images can ascribe social, political, and moral status to embryos.[32]

— Of particular interest to me is the manner in which visually depicting an embryo politicizes it and makes it "known." How is the embryo's meaning (and status) variously conferred or implied by direct sight, drawings, ultrasounds, paintings, photographs, sculptures, virtual models, in-utero video, or MRIs? The embryo is considered in our culture as property, family member, medical cure, bearer of legal rights, experimental material, and a potential—and an actualized—individual. These new depictions leverage the technology of MRI to inscribe views of human embryos that conflate the ideas of familial relationships, history, genetic manipulations, and human identity.

[public blog]

SEMEIOTICA

— I wanted to respond briefly to the first and third provocations: "What role do picturing practices play in your discipline of 'knowledge production'?" and "What role have new imaging technologies played in your conceptualizations of visual modeling or artistic application?"

— One outcome of creating images, maps, and animations is that they start to push and tug at existing knowledge by revealing gaps in what we know, or think we know. In the OrganelleView project we sought to create a visual, spatialized representation of a yeast cell from a very abstracted set of verbal data. We used a dataset along with artistic renderings and relational aesthetic methods to merge the identity and locations of gene/protein products with their localizations in a cell. The idea had two goals. By replacing the verbal data with visual cues, we could create additional means for exploration and discovery–particularly for those not versed in the specialized vocabulary of gene/protein nomenclature. Second, we wanted to reestablish a cognitive, visuospatial link to the organism (*Saccharomyces cerevisiae*), which is often abstracted through the collection and curation of biological information as text and numbers in databases.

— One of the more interesting outcomes of the project happened when we started thinking about animating the cell through the cell cycle. There isn't a great deal of organized information about where gene products locate at different times during the cycle. More important, the animation tweening process stimulated questions about the direction and paths that gene products move through during the cycle. What started as an animation question became bounded by questions of verifiability and if our animating techniques would accurately represent a version of cell processes. To my knowledge, current imaging techniques do not have the capability or feasibility to track these processes over time. Needless to say, we were impressed that "simple" picturing processes had the capacity to generate knowledge by highlighting missing data, scientific focus, and interest.

SUZANNE ANKER

— Dear Semeiotica,

— You talk about pictures as providing a "cognitive, visuospatial link to organisms" as way of filling in or providing direction to your laboratory investigations. Point well taken.

— Do you care to identify yourself?

— Dear Semeiotica,

GABRIEL HARP

— Hi Suzanne,

— Ha, sorry, I didn't even think to identify myself. Sometimes my posts get linked up to "Semeiotica" via my blog.

— Great forum, though. I'll probably be using this conversation for my students for years to come!

CATHERINE WALDBY

— Following on from Susan Squier's and Michael Sappol's points and in response to Suzanne's requests for more examples, Brad Smith's very touching and fascinating posting is a terrific example of what I was trying to get at (although not strictly an "art" context). The verisimilitude of digital imaging, combined with its spatial versatility, its novel forms of animation, etc., mean that these images invite speculations from both scientists and public viewers about the value and capacities of life, exactly because they allow vivacity to be modeled in new ways. As Brad says, his images allowed new types of biological questions to be asked and researched, while also provoking public viewers to ask what this new biology means for them. Both types of questions are essentially fantasies, but as I've argued in a few publications,[33] biomedicine and the life sciences have rich imaginative practices and need fantasy sites (e.g., the VHP [visible human project], the embryo images) where researchers can have access to an imaginative resource, especially a public one that a field can hold in common.

ORON CATTS

— Hi all,

— Need to be quite telegraphic today, as we are preparing for a living installation tomorrow; interestingly enough, the installation is in a political context that explores issues concerning artists transgressing the boundaries of art and other practices.

— Working for the last eleven years with the use of tissue technologies as a medium for artistic expression, I moved away from the notion of picturing what we do (representation) to showing what we do (presentation, and in some cases documentation). I also (in a somewhat simplistic way) do not believe that as an artist working in this area I am involved in "knowledge production," but rather, I am involved in (sometimes futile) exercise of "meaning production." The new knowledge I engage with (as you will see below) is mostly already there; the question may be: What does this knowledge mean to the rest of us? Artists use different strategies to deal with the very radical shifts in our perceptions of life, which are presented to us through the life sciences and their technological applications. Catharine mentioned Patricia Piccinini, who chooses to fabricate fantastical scenarios. That in a sense, can only come through the use of (mis)representation techniques, while my own practice deals with the direct use and engagement with the tools and techniques of a specific branch of the life sciences. What I presume we share (with many other artists in this field) is a desire to make sense of what we are confronted with by making it strange.

— As mentioned above, much of what I find challenging is the radical shifts in what we are coming to know about life through science and, more importantly, the radical actualities that are being executed on life through the way this knowledge is being applied (technology). Working with fragments of complex organisms required me to reassess constantly my perceptions of the body, life, and its classifications. Even though I work with "science," I see my work more to do with life then with art-sci. Actually, I am quite concerned about much of the rhetoric surrounding the art-sci arenas, but I will keep my thoughts about this issue for a later stage of the symposium.

ANDREW CARNIE

— Ha, let's get going. I must rise to Suzanne's earlier challenge to write something and see what happens.

— I feel my role as a practitioner, as an artist dealing with scientific topics is to add to the debate, throw open a space to think about the issues. This means having at least some knowledge of the area of science I am looking at, in order to be embedded in it. This has normally meant working quite closely with scientists in the discipline I am, at any one time, looking at. With Dr. Richard Wingate, a neuroscientist at the MRC [Medical Research Center] Centre for Developmental Neurology, Kings College London, I was able to extract visual information for *Magic Forest*, a light projection sculptural installation.[34]

— When I finally make the "art" work, I always back off from the science to give myself a little distance. I am not trying to make a descriptive piece about the science, but to convey some other "sense" of the science in terms of what it might mean to us psychologically, philosophically, or socially.

— As such, I don't think I am involved in fundamental "knowledge production." I might be involved in the implications and implementation of this knowledge for society at large. I think with the right applications, artists could be more involved in the lab by employing their strong visual and interpretative skills to the laboratory experience.

— In delving into the field of developmental neurology, I have been struck by the new images and research that give a sense of the changing body. I remember, as a schoolboy some forty years ago, being taught that the brain was a fixed structure developing to fruition at the age of sixteen or seventeen. After that, it was all decline. New fMRI [functional magnetic resonance imaging] scanning technologies change this, and, in a way, both my works, *Complex Brain* and *Magic Forest*, were responses to the ever-changing brain. The dialogue I have had with science has changed how I think about my mind and how I can use my mind creatively.

— Images screened from new imaging technologies are often amazing and overwhelming. They tell us about our bodies and our internal workings. The movement to less invasive imaging and imaging that can record over time is highly significant. The wondrous images produced by these technologies rivals images that artists once produced and brought back from exotic lands. The danger, as I see it, is for artists to use these images as a straight transcription into works of art, making paintings from photographs.

[public blog]

M M

— "What role have new imaging technologies played in your conceptualizations of visual modeling or artistic application?"

— The largest concern with "making paintings out of pictures" is that it dilutes the provenance of the image, and subsequently, the images lose meaning.

— For the uninitiated, the images need mediation, and that is the core challenge for an artist working with scientific visual data.

— In other words, the challenge is how can we [best] convey mediation?

ANDREW CARNIE

— I think simple transcription is not very effective, as

you say. Some form of mediation has to take place. In the work *Magic Forest*, I was dealing with the scientist Richard Wingate's QuickTime movies of the neurons developing in chick brains in vitro and also the drawings of neurons by Santiago Ramon Y Cajal from his various studies of the anatomy of the brain. The idea was always to convey the ever-changing state of the brain.

— Though at times my animations look like the originals, all my work was hand-drawn on the computer, and what gives it the biggest shift away from the originals is the way the images are then projected using two projectors on opposite sides of three voile screens. The images become slightly 3-D with the projection onto the multiple screens, plus one is having one's perspective of the images continually changed. At one point the projector is pointing the images at you; you are at this point subsumed into the image. Further still, the voile screens move in any breeze caused by a passing person, bringing the images to life.

— You can see the arrangement of projection for *Magic Forest* at the web page for Slice.[35] The arrangement was exactly the same: two projectors, three voile screens, 162 slides, and a dissolve unit.

DOLORES HANGAN STEINMAN

— I was mesmerized by *The Magic Forest*, as well as by Wawwa, Slice, and Eye.TTMD. I felt attracted and intrigued. From the point of view of the scientist collecting laboratory and clinical data and then translating them into (visually appealing) computer-generated images, I feel it is essential to acknowledge the influence art has on the scientific image. It is also important to emphasize the difference between the perception of a "documentary" image ("simple recording" of an event, and, as such, involving no manipulation) as opposed to the ones we "construct" (real-life look-alikes that have been created from real-life data but involve two conversions: a) from data collected from the individual patient to numerical data; b) from numerical data thus obtained into computer-generated images.)

— Which one is more reliable, or closer to reality? What is the message the images are trying to convey?
What is the purpose, or final use, of these images?

— The bioscientific image is not merely mirroring

reality; it converts, deciphers the code and at the same time visualizes the invisible. It is a process of interpretation of reality. By exploring new frontiers, by rendering the invisible visible, are we creating our own narrative, parallel to reality? To quote Nicholas Mirzoeff, "The visualization of everyday life does not mean that we necessarily know what it is that we are seeing." Thus the question of the role played by imaging technologies is as acute in science as it is in art. Merriley Borell considers the instruments used in experimental medicine as "extension of the senses" of the medical and scientific observer. The new technologies appear to have taken us on a whimsical road that leads us closer to the artistic process than we would've imagined.

M.I.D. VAN RIJSINGEN

— Hello to all from Amsterdam!

— "Picturing practices" in my discipline (art history) is, as you probably know, "core business." The question of what role it plays in "knowledge production," however, is blurred in the course of its 200-year historiography. Art historians have talked about the image (and about image production) as a source of knowledge. But asking what kind of knowledge, you will get very different answers. I tend to think that this is sometimes not very different from how scientists talk about their images. However, the new generation of art historians is much more into the image and image production as a productive signification practice, considering the image not as a source of (some hidden) knowledge, but as a productive force. And with it the discipline becomes much more a discipline of knowledge production. Or let me say, it takes itself much more seriously. This "turn" (linguistic turned iconic) was in itself a boundary project without which we still would be blurred. For me this "turn" meant also a new opening to the perceptual side of things, specifically the embodiedness of perception.[36]

— My research projects have always been around the body, and representations of the body. I moved from semiotics to phenomenology (and back). Current dialogues, but specifically current (art) works, have focused my view much more (and again) on the problems of the embodied "reading" of "visuals" (not only images, but also objects, installations, etc.) and the ways in which these readings produce knowledge. In the art-sci arena, the real is framed from both fields. In many artworks – I find – it is this framing that could produce knowledge, in that it makes us edgy "readers."

— I hope I am not too thick here. Will elaborate later on probably, if required. Looking forward to reading more.

CARL DJERASSI

— At the risk of partially repeating a point I already made in my original comment and my response to Squier, I would like to point out how I have used visual imagery in an art historical content in my recent play *Phallacy [fig. 6]*. In the audiovisual component of the play, you see two images of a bronze phallus cast twice. They move towards each other until they virtually, but not totally, overlap, indicating that they cannot be identical casts *[fig. 7]*. Since my play deals with an actual, though little known, art historical controversy at one of Europe's largest museums including the contrasting behavioral professional idiosyncrasies of art historians and chemists, by using art historical images, I am able to effectively show difficult subject matter such as Jesus' erection as depicted in this woodcut *[fig. 8]*.

— Any participants of this symposium residing in or near New York can actually see the use of these and many other images and bronze scans during the North American premier of this play (May 15-June 10, 2007), at the Cherry Lane Theatre in New York City.

ROBIN MARANTZ HENIG

— My experience with pictures is a little different. I'm a science journalist, so my metier is words, and the pictures are added by others (and then only when my work appears in magazines, never in books). A few of the articles I wrote recently for the *New York Times Magazine* were illustrated with photographs by Nicholas Nixon, whose sensibilities about the human form must somehow be in harmony with mine.[37] These articles were about difficult topics – one was about our quest for a good death through hospice and palliative care; the other was about a progressive aging disease called progeria—and I thought that I approached them directly and honestly, just as Nick's stark black-and-white photos did *[fig. 9]*. But several readers told me that they were really put off by the photographs, and a few said they had even decided not to read the article because they were there. A friend who was dying of cancer saw my article about death and wanted to read it, but she specifically asked for a version of the piece that she could read that was text-only, without the disturbing photos. I think this reveals the emotional wallop of visual images, which far outstrips whatever it is we writers can do with words alone.

[public blog]

ADRIENNE KLEIN

— Our Science & the Arts series (http://web.gc.cuny.edu/sciart) at the Graduate Center of the City

University of New York recently presented a reading of selected scenes from Carl Djerassi's play *Phallacy*. Djerassi's article in *American Theatre* (http://www.djerassi.com/ScienceStage.html) includes this exchange from the play, between a chemist and an art historian in a feud over the correct attribution of an art object.

— REX (chemist): Unpredictability is what science is all about...

— REGINA (art historian): Is it really? And even if it is, then why doesn't that teach you humility...rather than arrogance? And why not recognize the importance of visual beauty...a concept that barely exists in your chemical world.

— If an historical object also possesses beauty, that is value added; indeed it may be the significant value of the object. Objective documentation of scientific observations can have beauty as an ally, as in the arrestingly beautiful images that illustrate science by Felice Frankel (http://web.mit.edu/felicef/). Her representations of science draw readers to the text. By contrast, science writing can, as Robin Marantz Henig notes, suffer by association with unsettling imagery.

— I will make a leap here between the assertion that unpleasant images turned Marantz Henig's reader off and the claim that Djerassi cites, that audiences go to the theater not to be educated but to be entertained. Djerassi offers the challenge, "why not use drama to smuggle (with a substantial dose of theatricality) important information generally not available on the stage into the minds of a general public?" Certainly theater has often been employed to tremendous effect in conveying political sentiments; why not science-related ideas as well? Djerassi is accomplishing this in his plays and his new initiative for science theater in the classroom.

CARL DJERASSI

— Thanks for the compliment, but the somewhat defensive nature of the quote from my *American Theatre* January 2007 article is really based on the all too frequent utterly dismissive attitude of nonscientific theater people claiming to do science concerning even the slightest whiff of didactic motivation.

— In a recent book review (*Physics Today*, Feb. 2007: 63-64), I quote some smart aleck wisecracks from the French theater director Jean-Francois Peyret, who has tickled the fancy of some academic theater scholars enamored by postmodernist approaches to science theater. Here are two examples:

— "(1) If audience members want to know whether Heisenberg was good or bad, they have access to the scientific debates...they don't have to see a play."

— "(2) We don't have to do night schoool...We do not do scientific theater; we in fact do not even know what that means."

— It is this kind of puerile braggadocio that unfortunately carries all too much weight in many theaters and thus influences also prospective playwrights.

MAX AGUILERA-HELLWEG

— I feel that my thoughts, feelings, and experience here echo that of the journalist Robin Marantz Henig. I am a photographer and then became a doctor, and have come back to taking pictures again. Prior to entering medical school, I'd photographed well over a hundred surgeries. When I first started out, I was on an assignment, but I returned because I had to. It was this incredible mystery I had to return to again and again. When I first began to show the work to people, I didn't realize that some would have trouble with it, and when I got the first comments of disgust, I was troubled, didn't know how to respond. But I continued, because I had to; however, after a while the difficult comments I received began to wear on me, and I began to doubt where I was going. The attacks were mean and vile. It took some time, but slowly I began to realize that I need to continue, not for the people who respected my work, who trusted what I was doing, who found it even beautiful, but for those who couldn't even open their eyes to it. I needed to continue for them.

— What I realized was that the work I was producing was like a Rorschach test, that people's responses said more about them than it did about me. The images confronted their sense of mortality, and for those who could not face this truth, their vision of a "good death" at risk, even shattered, I realized I needed to keep taking pictures for them, so that one day, they would have a library, a book of pictures, a "production of knowledge" if you will, that they could come to, if they wished, and slowly peel away their fear.

— As I entered medical school, learned anatomy on a host of cadavers who only weeks before had uttered their last breaths, learned medical science, and entered the wards as a medical student, I saw that fear firsthand, but this time it was real in the eyes of a man with terminal COPD [Chronic Obstructive Pulmonary Disease], who struggled for each breath the last thirty-six hours of his life. The fear in his eyes was palpable. Yet I did not have the experience to understand what I was seeing, the knowledge of his disease to know his death was eminent, or know I could have intervened. A morphine drip to help him breathe and ease his way out perhaps. I'll never forget his eyes. As months and years moved on, my experience, knowledge, and responsibility increasing, I gained an intern's knack to tell if someone was not going to leave the hospital and acted with all my skill, training, and resources to help those I could to make their way home. More than once I found myself on familiar territory—patients and families with an unnatural fear of death; even when death was mostly likely, the disgust was most vile. I made it my responsibility to be gentle, to teach them what they needed to know, to keep them informed, to empower them, to give them the facts to make the decisions that needed to be made, that maybe, just maybe, a "good death" might occur. Those who fought death had a wretched time. People still react as they always have to my photographs, some are fearful, others find their beauty.

RICHARD TWINE

— I'm a sociologist. Where imaging in science has come into my work is when I first looked at eighteenth / nineteenth-century physiognomy and phrenology. Here, of course, images of the face—often already in caricature—were endlessly reproduced to try and solidify the notion of a fixed relation between bodily exterior and psychical interior, and specifically more fairytale/ culturally entrenched discourses about beauty and goodness.[38] Similar claims on the truth-telling properties of the human body are now revived—as Anker & Nelkin also point out—in contemporary genomics discourses, yet now we move to the physical interior of the body, be that sequences of genomic code or the sort of imaging of the interior described well in Waldby's work. I have skepticism for the ability of such visual discourses to say anything particularly interesting or holistic about the "human," seemingly still positing "that" as a fixed property waiting to be revealed. I also have a developing interest in the visual practices of advertising work carried out by emergent biotech companies, and the visual practices of scientists in their PowerPoint presentations.[39]

GIOVANNI FRAZZETTO

— In my experience, the intersection between the scientific and the artistic practices constitutes an excellent platform to recognize the ways in which visions of science have become narratives embedded in cultural phenomena. Much as Susan Squier has pointed out, I attribute an enormous potential to the immediacy and succinctness of images in highlighting the dense mixture of social, political, and personal conflicts that can arise from contemporary scientific progress and practice. An educated scientific knowledge grants a more considered perception of the immediate benefits and dangers that can be drawn from scientific study. More important, it allows a balanced appreciation of the social and ethical issues raised by the new technologies. However, the interaction of art and science is established as an essential avenue for innovation and intervention, and as a way to explore, envision, and critique possible futures and societal aspects of scientific progress.

— Artistic representation and its imagery show how some of the seemingly intractable questions posed by the life sciences can be addressed: by exposing ambivalent emotions and provoking further reflection and discussion with irony and ambiguity, which science often cannot afford. Museums and art galleries are social spaces, and they afford an opportunity where scientific knowledge can meet the cultural imaginary.

— I was recently involved in the organization of a sci-art exhibition together with Suzanne Anker. The project was called *Neuroculture: Visual Art and the Brain*,[40] and it intended to emphasize how the mechanistic understanding of brain functioning is articulating a new vision of humanness not only equated to the genome but to the brain and its complex networks. In this vision, a more commonly cherished sense of the "self" as an individual inhabited by a deep internal space shaped by biography and subjective phenomenology is replaced by a somatic sense of ourselves, mapped onto the brain and susceptible to intervention and manipulation.

— For the project I divested myself of the garments of science and challenged, by the creation of an ironic work of art, the benefits of advances in psychopharmacology and the potential danger of an unbridled use of psychotropic drugs and of neurochemical enhancements. In the piece called *The Qualia Bar*, qualia, which are subjective states of sentience, acquire a synthetic and reproducible nature *[figs. 10, 11]*. The desired state of sentience is achieved through the intake of colored liquor. In an analogy to Huxley's *Brave New World* and the drug "soma," citizens and consumers live in a society where the fabrication of emotions is rendered possible.

— In line with what Andrew Carnie wrote, I would be inclined to say that the new imaging technologies are pervasively changing our perception of the body and consequently expanding the tools to formalize a perception of ourselves. The improving resolution of brain imaging techniques allows us to see the brain at work, in real time. Regardless of what these pictures translate into—and we can enter a debate on whether it is plausible to claim that we see emotions or states of mind when we watch the internal workings of the brain—the mere visualization of the cerebral activity is bound to leave us with a renewed perception of ourselves, impregnated with wonder and curiosity. I leave a more sophisticated interpretation of the leap in brain imaging technologies to an expert practitioner.

JILL SCOTT

— I don't really agree with the term "picturing practices," as I see my role both as an art and science context provider[41] and as an "art researcher" working in neuroscience[42] who is challenged by both the differences and similarities between visualization, illustration, simulation, and interpretation. Although many art researchers prefer to define their works as interpretations of scientific research, I actually think that the most controversial and reflective practices often sit in the fuzzy boundaries between the definitions of these terms. These boundaries are further blurred by the fact that reductionist methodologies in science (including psychology and neurology) are currently under great scrutiny. Meanwhile the art researchers (including me) are struggling to find alternative ways to question or be inspired by scientific inquiry or even construct a "proof of theory" rather than replicate empirical-based methods. Perhaps "knowledge production" in both art and science has been very culturally situated, but this may not be the case in the future!

— It seems that art researchers are led through scientific discourses based on their own trajectories rather than fashionable discourses. My interest has always been the human body. Once in 1995, I conducted a great deal of research and visited many labs for my practice-based Ph.D. about the impact of biotechnology on the human body.[43] Since then, my interest grew to include artificial intelligence (sensory perception), neuroscience (particularly cross-modal interaction), and environmental science (embodiment). Indeed this trajectory has led to big shifts in attitudes, one which should reflect the impact of science on society and the need to create technologies that can actually help people.

— New imaging technologies have played a great role in my own practice, including the construction of interactive interfaces for the public to unravel linked narratives. As I was interested in interactive media, I often simulated cognitive relations with 3-D animation. In the last three years, visual modeling has played an important role for me in designing mobile prosthetics for visually handicapped people. Currently, I have a commission for a lab based in neuromorphology (Uni Zurich) to build a mediated sculpture. This lab gives me hands-on access to many visualization methods (i.e., SEM [scanning electron microscope]), that produce exceedingly beautiful images. This beauty is in itself very problematic and a deep challenge for any artist *[figs. 12, 13, 14, 15]*.

ANDREW CARNIE

— Jill, I am very interested in what you have to say in your comments and wonder if you could elaborate on the issue of beauty in art, especially relating it to the science area. You say, "This beauty is in itself very problematic and a deep challenge for any artist."

— In what way is this true? How do you overcome it yourself? Are there art market pressures to conform to a type of beauty or aesthetic? How inbuilt is this aesthetic to every human, or is it cultural? What are the particular pitfalls in the art-sci/sci-art arena?

JILL SCOTT

— In my original quote I said, "The beauty of scientific images (like those from the SEM) is in itself very problematic and a deep challenge for any artist." From my experience so far, there are two problems here. First, natural scientists have a rather old-fashioned view of the relation between beauty and art, and they often perceive their own images (e.g., nanoscale images) as art. Second, artists feel intimidated by these images, as they are so seductive and, in a way, sacred. It is difficult to imagine using them in an art context. In other words, they may not inspire an interpretation because they are such strong structural statements in themselves about a certain part of the body or plant or substance. These images are often the essential ingredients of that object.

— Currently I am trying to build an inverted retina (a commission for a neuroscience lab), and I am hoping to abstract essential forms and shapes from microscopic images to build the mediated sculpture. So abstraction is one way to deal with the problem. But there must be other ways, and I would hope that further brainstorming could help artists in this regard.

— You aked if there are art market pressures to conform to a type of beauty or aesthetic? Yes and no! I have to say I am really not concerned with the art market; I would prefer to have an audience who is interested to learn and think.

— "How inbuilt is aesthetic experience to every human, or is it cultural?" Our primal reaction to beauty is absolutely a genetic trait; we all respond in similar ways to sunsets or to micro images of our own heart cells beating. Cultural influences sway interpretations, so we really have to care about them, too.

— "What are the particular pitfalls in the art-sci/sci-art arena?" I have a long list of problems about this issue, but I will present them in the form of questions. So perhaps you could suggest which ones you would like us to discuss, and

I hope others may also join in. Perhaps they can be reposted to the others.

— Here they are:

1. How necessary are disciplinary hierarchies for art and science researchers?
2. How can we promote know-how transfer between art and science?
3. Can methods or methodologies be shared?
4. What do the terms "radical" and "ethical" mean in art compared to science? Is there a concept of value-free science?
5. Can consideration of place, community, or culture affect scientific search for empirical knowledge?
6. Do art and science share similar attitudes towards creativity and innovation? Do differences in these definitions have any impact on gender and representation?
7. How do artists relate to terms like "technical progress" and "information society"? Does capitalist ideology help science and their related businesses, or does it hinder production or deter progress generally?
8. If artists share more poetic metaphors with scientists, will the results be more suitable for tele-visual literate societies?
9. How does the public benefit from art embedded with scientifically robust knowledge? Could psychological evaluation help art and science to communicate more clearly to the public?

ANDREW CARNIE

— "How does the public benefit from art embedded with scientifically robust knowledge?"

— The former, I think, benefits from a "rounding," a fuller, more conversant communication of the science work, though, of course, one also wants to be able to "expose" the science for what it is sometimes, contextualizing some of the negative aspects. This might lead to a slightly difficult, less embedded relationship. Certainly in my projects I want to move away from the science when I make the work. I want to stand at some distance from the science and maybe only touch on it slightly, and concentrate more on a human dimension. This is also to avoid simply repeating the science and sometimes because the science is just too complex to make work about. I once had a fascinating contact with the Medical Research Council Centre for the Study of the Synapse in Bristol, UK. The work done there was amazing but chemically so complex I just couldn't find a way to engage with it artistically. The work did affect, though, how I made a subsequent piece, though I could not deal with the content of the research group.

— The latter part of issue nine suggests to me that you think more linkage, more embedding of artists and scientists, maybe even more training, would be a good thing. I once took part in a collaboration with a dance company to make a combined work. It was not an easy relationship en route to, though, what turned out to be a positive result. A third party afterwards suggested that what would have been a good idea would have been for both parties to attend training sessions in collaboration before undertaking the venture. I agreed with this; some instruction to see where the common ground lay and where differences in expectation were located and the provision of processes of resolution would have been good.

JILL SCOTT

— In our AIL [artists-in-labs[44]] program, we actually pay the scientists to teach the artist some basic foundations of the scientific research at hand. Perhaps these are some other relevant points or questions for this debate.

— Is it interesting for you to reflect on the issue of the public being properly pre-exposed to the role of scientific research and resultant proofs in order to explore the prospect of the public value of scientific research itself?

— I also wonder if art should only be a catalyst for science to reach the public, or should it maintain a purely interpretative role? In the AIL project we hoped that matched pairs of art and science researchers who collaborate on very particular problems could help to promote the production of artworks that are based on scientifically robust knowledge.

— We also wondered if the intelligence of the audience can be more taken into account. This is a very different approach to commissioned illustrations of physical or biological principles in order to enhance the general public's comprehension and understanding of these sciences. As Donna Haraway[45] suggests, if science is really neutral, then artistic interpretations might also reflect how cultural factors, through feedback from the public, might have an influence on science research itself.

— I actually think we should encourage artists to become more serious researchers. This is in accordance with [Silvia] Caravita's [46] idea that knowledge building "receives promotion from the cultural environment and the social interactions that accompany the learner's explorations."

— Please let me know what you think?

ANDREW CARNIE

— Dear Jill,

— What I had in mind was that one actually needed training to undertake collaboration, that both parties, scientists and artists alike, needed to understand better the nature of collaboration, what it might mean and how to negotiate the pitfalls. Then after this, you are right, both parties need training on what the respective domains of each area are. Then you can get into the experience of what the particular areas of each discipline are and the exchange of specific knowledge.

A three-stage process at the least?

V L A D I M I R   M I R O N O V

— We had a similar problem in tissue engineering at the beginning. The question was, "Do we need to teach engineering to biologists or biology to engineers?" The answer was that biologists and engineers must jointly teach a new generation of scientists a new evolving profession - tissue engineering. Maybe artists and bioscientists can jointly teach the bio-artist. Learning of anatomy and anatomic drawing is already an integral part of classic artist training. Who knows, maybe a bio-artist training program based on seamless integration art and biology training will be very popular and successful. At least from my perspective this is definitely something to explore. It would be interesting to know if there is any demand or room for such specialists in the art and entertainment industry.

E U G E N E   T H A C K E R

— I definitely agree that context matters a great deal. I'd be interested to hear from others who've dealt with similar experiences (Jill, perhaps you can also talk about your work with Z-Node [Zurich Node of Plymouth University, Zurich]?). Being at a tech university, I've found the art-sci thing to be very erratic. It can be really easy to have lunch with a colleague from the Biomedical Engineering school, but quite a different matter to collaboratively apply for a grant or work on a project or, for that matter, to co-teach a course. Often it boils down to institutional framing, disciplinary flexibility, and budgets. A colleague of mine in the Biomedical Engineering actually did collaborate with SymbioticA a few years back (his rat neurons controlled a robotic arm over the Internet that made drawings). I went to visit him once, and, in passing by his lab, I saw that he put up a huge poster/publication of his work with SymbioticA. I really respected that. It was a little gesture, yes, but it was saying "this is my research" (almost every other lab was working on NIH [National Institute of Health]-funded stem cell research). Of course, I think he has tenure and plenty of prior, um, "legitimate" research.

— I find the institutional context for this kind of collaboration to be at once intriguing and frustrating. There's a lot of double-talk. On the one hand, you can create new degree programs and hire innovative faculty to teach in those programs. But on the other hand, you're given strict guidelines from above on what is or is not "tenurable." So you can create your crazy art installation as long as you publish that scholarly monograph. But many artists simply don't self-identify as theorists or writers. I've seen many media and visual artists leave their jobs because this double-talk has imploded. Sorry to raise such an unexciting issue!

— Here is my brief stab at opening answers, during the latter end of the flu. Such is the deluge of material already, I fear I will never catch up. There is no chance of my reading all the contributions given my schedule over the next weeks. Sorry. I am a kind of visual historian, paid to teach history of art but also engaged with images from the worlds of science and technology (see my monthly column in *Nature*).

— Pictures / images generated by others are the subject of my analyses. The analyses are a form of directed pointing, aimed at producing insights and fresh perceptions of the images, above all (since I am a historian) in their contexts; a hard and necessarily selective job, not least because the contexts don't define themselves. It is, of course, possible (as much current art history vividly demonstrates) to produce seductive pseudo-insights that have more to do with the state of the subject in academia rather than with the period generation and reception of the images.

— The tools I use are both traditional (verbal analyses and word-painting) and modern (see below). In my analysis of the relationship of optics to Western painting, I have used drawn diagrams.[47] I still prefer to conduct such analyses by hand, not least because manual drawing bears some relationship (if conducted on as large a scale as possible) to the artists' actual procedures (but now see computer vision, referenced below). Much of my research, teaching, and publication concerns knowledge generation through the history of images from the Renaissance onward. I am concerned with the relationship between claims of veridical naturalism and what I have called the "rhetorics of reality." I am not using the term "rhetoric" in the now-fashionable sense that rhetoric equals false. Rhetoric was, after all, "the art of persuasion" and for Cicero involved the "invention of something that is true or plausible." Strategies of visual persuasion, often based on what Gombrich called the "eyewitness principle," can be laid bare, but it does not mean that the subject of the strategy has no truth content (however defined). The same basic point applies to machine-generated images, which are often suffused by the rhetoric of high-tech.

— Having been involved in the so-called art-sci/sci-art territory for forty years—from a time when it did not exist—the burgeoning art production, the contributions of authors, scientists, historians, and commentators have provided the ever-changing context within which I have worked. My involvement with artists I regard as vital in refreshing perceptions and even in altering conceptual fields. The historian on one hand has to avoid simply refracting the past through the present, while on the other adopting fresh perspectives and modes of interrogation that allow the past to speak to us in new and sometimes unexpected ways.

— New imaging technologies have played a significant role in two respects: research and public exposition (though there is, in practice, no clear separation between them). In the Leonardo show at the Hayward Gallery in London in 1981, we worked with the IMB Research Centre in Winchester to animate some of Leonardo's ideas (geometry, optics, perspective, branching systems, architectural design). This was undertaken in the conviction that the "architecture" of computers could be placed in creative dialogue with Leonardo's ways of thinking visually. Inevitably, generating the input for IBM and responding to queries resulted in having to ask new questions (or answering questions that I had neglected to answer). The same applies to the rather different types of animations used in the recent Leonardo show at the Victoria and Albert Museum.[48] The new animations were based on the inherent sense of spatial movement in Leonardo's drawings—what he called "continuous quantity." I've called this "scratching itches that Leonardo could not scratch."

— Another strategy has been my collaboration with Antonio Criminisi, specialist in computer vision (formerly Oxford, now Microsoft Cambridge). We have used his technique of "single-view metrology" to undertake direct analyses of space in paintings (no extraction of data and insertion into CAD [computer-aided design] programs). More recently, we have "rectified" the convex mirrors in paintings by Campin and van Eyck, with astonishing results that demonstrate how Netherlandish artists used "actors" in real settings to "stage" their astonishing feats of naturalism.[49]

— The new techniques are not more "scientifically objective" than the old ones. Rubbish questions, misapplication of techniques, and wishful thinking still generate rubbish. But they do provide potentially powerful tools to answer old questions and bring new questions into the frame.

— Not much of this is to do with bioscience. But you did ask.

CATHERINE CHALMERS

— Greetings from the Liquid Jungle Lab on Isla Canales de Tierra, Panama.

— I am here working with leaf-cutter ants. Or maybe it is the reverse—they are working on me. Through observation and experiment, the animals I film and photograph slowly impart their "knowledge." And it is this "knowledge" in relationship to our "knowledge" that interests me.

— This is the first time I have worked in the field with a so-called wild animal. Prior to this project, I had raised and filmed the animals in my NYC studio. Last night (during this season leaf-cutter ants are mostly nocturnal, and then sometimes only part of the colony, hard to figure this all out—any entomologists out there?) the howler monkeys came to investigate. At first, from a distance, they were screaming. But when they came close, right above our heads, their voices became soft. They sounded more curious than challenging. It was quite a moment. Perhaps our strobe lights brought them in.

— I often consult with scientists to learn the basic aspects of an animal's behavior. For example, when I photographed Periplaneta americana having sex, it was vitally important to know (well, most obviously, how to distinguish a male from a female, juvenile from adult) that they are capable of parthenogenesis (if no "attractive" males around, the females can lay clones), and that once the females do choose a mate (they're very picky), they mate once and are pregnant for life. Who would imagine the lowly roach to have discerning taste?

— In response to the three questions—as a visual artist, pictures are the only means I have of "knowledge production." And advances in imaging technologies—every gain in photographic and video quality, clarity, editing, and printing—make my exploration of the animal world that much more rewarding.

— I may lose internet connection at times, but will try to keep in contact.

SUZANNE ANKER

— Further Thoughts on this Moveable Feast of Ideas:

— How do pictures embedded within texts alter narratives? What value-added aspects do pictures "perform" in texts? When do pictures become value-subtractive? This is a point raised by Robin Marantz Henig, yesterday. Roland Barthes, in his "Rhetoric of the Image" talks about the ways in which language, anchored to pictures, is an effective strategy in advertising.[50] The language in the ad grounds the image to further enforce and relay its message. Richard Twine speaks of his interest in these commercial visualizations. However, a variegate host of types of images exists on all levels of inscription, documentation, and communication. Do not progress notes, letters of inquiry, travel stubs, and the like all become mimetic devices embedded within critical thinking processes?

— How do working models intersect linguistic and iconic metaphors? In my own practice, images play a dual role. They are a thinking apparatus by which models of the symbolic can be externally visualized. By their existence, these models move into culture and generate, on their own terms, alternate ideas, solutions, concerns, and further representations.

— Perhaps no other discourse is so indebted to images as art history. Recently the term "picture science" has surfaced as a way to explain more culturally based narratives involving perception and notation. How can the terms "picture science" and "artist/researcher" be communicated to a wider audience? What do they mean?

— Materials and actions also leave traces as they are manipulated in the laboratory, studio, or writer's abode and could be construed as imaging processes. These indexical signs denote how thinking becomes form and carries within them the unwritten script of trial and error. Image-making here and in all my questions and responses can be conceived as what the late American philosopher Nelson Goodman terms "world-making." [51]

VLADIMIR MIRONOV

— Picturing practices is the essence of anatomy as a morphological discipline. Real microscopic pictures as well as explanatory schemes and animations are very important for presenting anatomical and cell biological knowledge and scientific visual information. In this case, seeing is believing, or a way of objective documentation and presentation of scientific knowledge.

— Because up to now I was not directly involved in art-sci dialogues, I could not say that my perceptions and attitudes have been challenged. I believe that art and science belong to the same domain—human culture. Both art and science are idea-driven cultural activities.

— Visualization and new imaging technologies such as MRI and other clinical bioimaging technologies are very important in bioprinting technology.[52] Computer-aided design, or "blueprints," for organ bioprinting is basically visual modeling based on modified MRI images. It is not possible to build a bridge without a blueprint. Similarly, it is not possible to print an organ without visual modeling or computer-aided designs or blueprints for organ printing.

TROY DUSTER

— Greetings to all!

I have just returned from Europe, and am sitting in a transit lounge at Dulles airport en route to yet another destination.

— Since my recent interests have centered on the sociology of science, I will probably have much more to say in the third session, but here is a start:

— I am fascinated by the relationship between pictoral image and scientific authority, and depending upon the domain of inquiry, that authority can be placed along a continuum from "nearly definitive" to "merely anecdotal." In recent work in the neurosciences, for example, brain imaging technology permits certain claimants to assert, with extraordinary authority, that the way the brain circuits seem to "light up" in response to selected stimuli provides evidence for some complicated neurological process (e.g., racial prejudice when viewing images of members of different racial categories). The conclusion reached in some of these circles is that such attitudes must be "hard wired." Of course, this bypasses the prior question about how those wires got to be that way—but on the matter of authority, it is precisely this bypass that captures so much attention (as knowledge production?). Upon closer inspection, the various colors assigned to different images of how the brain circuits are firing is a decision by the researcher, and NOT how the brain "lights up."

— At the other end of that continuum is the relative lack of authority of the pictorial image of social processes. The more textured, detailed, and nuanced, the more specific to time and place and manner—thus rendering the images vulnerable to being dismissed as "merely anecdotal" in the jockeying for positional authority from skeptics or counter claimants.

— That will have to do for now.

SUZANNE ANKER

— Troy Duster's recent post brings up the nagging question of how new technologies masquerade issues concerning social injustice. Repackaging, as we are all well aware, has turned spaghetti into pasta, the garbage dump into a recycling center, and secretaries into administrative assistants, to name of few. The more pernicious issues at hand are the subtle interpretative factors that determine what is being "seen" and what, in fact, these instrumentalized images mean. A scurilous case in point is the rise of "brain fingerprinting" technologies as they are employed to provide evidence in criminal cases. Any neuroscientists care to respond?

SUSAN SQUIER

— Troy's post made me think about some images I've been looking at, and thinking a lot about, recently: photographs of chickens. Because I'm writing a book on poultry science and chicken culture, I've spent time looking at three types of chicken photographs: 1) breed photos, the kind of idealized images of specific chicken breeds that have long been important to purebred chicken raising, and that appear in the *Standard of Perfection* book on breeds, as well as on posters from feed stores; 2) photographs of chickens either taken by, or taken with, the farmers that raise them (for example, a photo I have, circa 1910, of a woman sitting in a long black dress holding on her lap two chickens that she has raised); 3) photographs of chickens taken by professional photographers (including the photos in Stephen Armitage's book *Extraordinary Chickens*, among other such photographs.) So: what relationship exists between each image and scientific authority? The first category is pretty clear: the eugenic gaze of the photographer situates authority in the role of the purebred chicken breeder, and secondarily in the eye of the judge at the chicken exhibition. In the second category, authority (particularly scientific authority) is more amorphous: the farmer has a kind of authority over the chickens she/he raises, and yet in the early days of the twentieth century that authority was arguably not specifically scientific. Of course, with the rise of the discipline of poultry science in the '20s and '30s in the USA, a kind of scientific authority increasingly came to the fore. In the third kind of chicken photograph, there is the scientific authority of photographic process from questions of aperture and film speed to questions of exposure and printing. But what interests me most about that third category is the way other kinds of disciplinary (or compositional / epistemological) authority intervene there.

— I am thinking in particular of a show by photographer Jean Pagliuso, up last year at the Marlborough Gallery in NYC, and now up currently (I believe) in Madrid, called *Poultry Suite*. These exquisite black-and-white photographs of chickens, in a larger-than-life format, printed on handmade Thai mulberry paper, invoke a number of different modes of authority simultaneously. Many photos recall exhibition breed photographs; one even recalls the psychiatric photography supplementing Charcot's work in the Saltpetriere...and yet when I interviewed the photographer, she found the most explicit relationship between photographer and image that of fashion photography. In addition to drawing on the grammar and techniques of the first two modes of photography, in other words, she explicitly found herself drawing on the grammar, design techniques, and even the vocabulary of fashion photography. I will try to post some of Pagliuso's prints here (or at least links to them); my interview with Jean Pagliuso is forthcoming *[fig. 16]*.

— Best to all, and now on to the Graduate Program committee meeting!

MIRIAM VAN RIJSINGEN

— When you mentioned your chicken project, the way you question the different authorities attached to different chicken photographing practices, I thought you missed the work of Koen VanMechelen, the Belgian artist who cross-breeds chickens to produce the "Super-Bastard" in his Cosmopolitan Chicken Project.[53] Part of his project is photographing his chickens, but also drawing them, making objects from them, eating them, etc. I would be interested what kind of authority is revealed in his work.

GIOVANNI FRAZZETTO

— I would like to followup on Troy Duster's and Suzanne Anker's interventions on "authority" and "repackaging" and will again focus on the neurosciences. In trying to understand (and question) the traffic of brain imaging evidence between the laboratory practice, the world of art, and the social context, I wonder whether it is legitimate to use and remind us (maybe rather hazardously) of a theoretical notion, the "Immutable Mobile," introduced by Bruno Latour twenty years ago.[54] In "Drawing Things Together," Latour asks: "How can distant or foreign places and times be gathered in one place in a form that allows all the places and times to be presented at once, and which allows orders to move back to where they came from?" He is addressing inscriptions of scientific practices and "worlds" onto objects, and, in our discussion we can say, onto "images." These images can be seen as representation conglomerates in which many worlds, practices, and definitions are reassembled and made presentable in a different setting. Paradoxically, these inscriptions are not in and of themselves explanatory, but they become so when considered mobile.

— Immutable mobiles, as representational devices, allow scientific practices to be transported from one laboratory to another and from one field to another. In brain imaging studies, the colors, symbols, and codes routinely and conventionally assigned to brain circuits are chosen by scientists to render the inscriptions presentable across laboratories. Those codes are, as Troy Duster has pointed out, not a direct reflection of how the brain lights up. In the laboratory, fMRI images enjoy an authority based on reproducibility and, for instance, from the juxtaposed representation of individual brain scans produced in one scanning session that, when assembled, become increasingly significant. When these images exit the realm of the laboratory and get embedded into context, they are laden with a different valence. I agree that their authority changes. And so does their immutability. If we think about it, these products are (or increasingly so) easy to produce. In daily life, anyone could in principle decide to have their brain scanned or to volunteer in a research study. Brain imaging pictures get easily transferred from a "scientific" world to another world, and they assume a power in determining subjectivity, justice, and responsibility to name a few. Their authority becomes pervasive and, I would say, reifying.

NANCY PRINCENTHAL

— What a fascinating conference so far. The frontiers of medical and scientific imaging do certainly offer a new landscape to visual artists, and also reconfigure the art audience. An important question raised several times concerns presentation venues and audiences.

— I'd like to introduce a question about new developments in the understanding of unconscious processes, as they relate to both the reception and production of visual imagery. It's clear we respond to so many visual stimuli without (or before) conscious awareness. Does anyone know of neuroscientific work on this issue that directly engages the visual arts?

— Also, does anyone know of neuroscientific work that provides recent (even speculative) findings about how the visual contents of dreams are produced? Is visual coherence a function of recall, or of the primary dream experience?

DAVID FREEDBERG

— Along with Pietro Perona, Vittorio Gallese,[55] and Fortunato Battaglia, I have been working for the last decade or so on the neuroscientific aspects of automatic and unconscious responses to visual imagery. We've concentrated particularly on the problem of emotional and motor responses. The degree to which these are precognitive or not has been a fundamental issue in the many discussions I've been having with both neuroscientists and cognitive psychologists, as well as in the experiments we're currently devising. Much of the work focuses on issues of embodied simulation. An essay by Gallese and myself which deals, inter alia, with aspects of automatic empathetic responses will shortly appear in *Trends in Cognitive Science*. [56]

RICHARD WINGATE

— Hi,

— By way of a brief introduction, I'm a research scientist studying the development of the brain. I've had collaborations in the visual arts (Andrew Carnie, a fellow panelist here) and history of imaging (Marius Kwint at Oxford University). I'm guessing/hoping that I'll be able to offer a slightly different perspective on visual culture and bioscience rooted within the lab—we'll see.

— Pictorial representation is of central importance within my area of science. Nothing, it seems, is quite as compelling as the elaborate images of individual nerve cells, the patterns of connections between brain regions, or the patchwork mosaic of transient gene activation during brain development. Images are often primary data, but they are also used for illustration or emphasis—a visual assertion. Quite frequently, a particular image or graphical style of interpretation becomes emblematic of a particular research program or lab. Pictures are therefore both the substance of scientific research but also convey a sense of style.

— Art-sci challenges the relatively easy assumptions of the autonomy of science from a set of broader influences. From the earliest stage of collaboration with Andrew Carnie, it was clear that the themes of my research (fate, commitment, lineage, guidance, migration, with respect to cells) were those issues that Andrew Carnie explored in his work. How much of the resonance associated with such words do we exploit in science, unconsciously or consciously, to add motivations and a sense of meaning to events under the microscope?

— Working in collaboration also raised the question of which biomedical images, within a given era, are chosen to represent science. Should we as scientists be concerned with preparing, rendering, coloring microscopic images for ubiquitous competitions or awards? Why do certain representations of biological constants sit more easily within a particular age than another? This kind of thinking prompted a more academic investigation of the history of the image of the neuron within its cultural context.[57] Again the art-sci world is the arena in which to reflect on what aspects of a broader visual culture we scientists pull into the pictorial representation of our data and its graphical, often schematic, explanation.

— Digital imaging has transformed biomedical science. Many scientists are experts in image acquisition and manipulation, and frequently write their own software solutions, where commercial products lag behind the pace of storage, rendering, and visualization requirements. For the last ten years, I have been working at least in some part of my research with 3-D time-lapse images, bringing me into contact with obscurities of MPEG [moving pictures expert group] encoding, film editing, and data file structure. [58] By contrast, in the not so distant past, much of the data I used was constructed by pen and ink. What's been an interesting revelation is that the unconscious understanding of structure that comes with drawing has been lost within this new digital, pseudo-colored world. I frequently find myself reducing these epic quantities of hard-won data to sketches on the back of an envelope—not only to explain to others, but also to understand.

JILL SCOTT

— Richard, I read your posting with interest and I have some questions. I would love to have your response.

— 1. Could you imagine that the use of more poetic visual metaphors by artists may help to translate certain perspectives of your research to the general public? Or do you think that public images in neuroscience journals always have to be literal representations?

2. Do you think that the actual process of working with an artist helped you to think "out of the box," and if so, in what way did it actually affect your "sense of meaning about events under the microscope"?

3. In many neuroscience labs behavioral attributes in

relation to neural research are drawn by trained scientific visualization experts, and their drawings often appear in PowerPoint presentations in conferences. What is the difference for you between interpretation, visualization, illustration, and simulation? Do interpretations only help to understand research processes when the artist has a very solid understanding of the research?

RICHARD WINGATE

— Hi Jill—and thanks for the stimulating questions. I will do my best!

— I think that Andrew Carnie has already had some interesting thoughts on points 1 and 3.

1. "Could you imagine that the use of more poetic visual metaphors...?"

— We tend to reserve our more poetical interpretations/rendering of primary data for specific situations such as the journal front cover, where images are often explicitly chosen on the basis of an uncertain aesthetic perspective (beautiful, humorous, pun-ish?).

— Within the scientific presentation (article, seminar) itself, schematic representations offer most obvious license for applying a design grammar that can be more or less effective and give a study more or less weight.

— But even with literal data, the choices that we make in rendering, thresholding, and false-coloring of images imply a poetic consideration. Certain styles of image presentation gain currency if associated with scientific success, and propagate accordingly: evidence that literal is a questionable quality.

2. "Do you think that the actual process of working with an artist...?"

— Rather than out of the box, working alongside an artist has helped define the box—the boundaries between scientific interpretation and a kind of metaphorical interpretation.

3. "In many neuroscience labs behavioral attributes in relation to neural research are drawn by trained scientific visualization experts..."

— My personal experience of having my images re-worked by professional journal artists is that they are unsatisfying because of a lack of authenticity in understanding a mechanism or structure. Maybe this comes back to my training, where visualization and interpretation were encapsulated in the process of creating images. I also think that it is possible to detect a lack of authenticity in illustration and simulation.

— So if authenticity comes from a proximity to visualization and interpretation, then one contribution that art-sci can make is to embed the artist in the lab and give the scientist an alternative vision of our models, literal data, and conclusions. To return to point two, the two interpretations may well, or should, remain autonomous but, ideally (for me), cast that perspective on the cultural load that we all bring to perception and interpretation.

JILL SCOTT

— I agree that literal visualization evidence is a questionable quality. But I also think that different approaches to communication often help or hinder the understanding of research. From our experience, one of the main differences between art and science researchers is in their use of metaphors. This concerns the value of visual poetic metaphors compared to a more literal use of metaphors. Often scientific researchers have not had any formal training in the development of visual metaphors, and educators in these fields tend to use language metaphors because they feel that they minimize ambiguity and seem to be understood by most people. However, language in itself can be very ambiguous and full of clichés and triviality, and these metaphors can cause conservative judgments and really affect levels of respect.

— This is an issue that may need to be taken into account in relation to the future of art and science collaboration, because I have the impression the basic scientific researchers are ambiguous about the value of metaphors. A literal interpretation of a metaphor can easily lead to a wrong understanding of the subject at hand. Some examples include: a visual metaphor of an atom as a solar system consisting of a sun (the nucleus) and its planets (the electrons)[59] or how biological evolution corresponds to the metaphor of climbing a mountain in a three-dimensional fitness landscape.[60] This lack of training and consequent misrepresentation in visual metaphors seems ridiculous when some of the most complex and beautiful metaphors can be found in science.

— Instead of using metaphors based on generalizations and language, the contemporary poetic metaphor is based more on thought and conceptual associations. As Lakoff suggests in *The Contemporary Theory of Metaphor*,[61] artists and writers learn that there are many conceptual types of visual poetic metaphors. There are structural metaphors related to the concept of dimension, whose dimensional associations may change with differing cultures. Another type of metaphor is about orientation and occurs when structures are experienced in terms of spatial orientation. A third might be ontological metaphors, which occur when our experiences are related to abstract phenomena or in terms of concrete textures, forces, or objects.

— It would be interesting to see what would happen if scientists were to start to explore more poetic metaphors, rather than those found on the front cover of *Nature*. What do your think about this?

— Could more discussions about metaphors between artists and scientists actually contribute to a highly skilled, critical, and reflective artwork, which might gain more respect from science?

SUZANNE ANKER

— Pictures thought to be grotesque usually contain disturbingly graphic body anomalies, both congenital and received. Sometimes violently criminal and other times the result of nature's endless bank of variegated form, these images indelibly mark our sense of mortality. Images of death and disease, body parts and transplants, surgical procedures and even birthing processes present the body as an envelope of painful eruption. To what extent have these uncomfortable images been erased from public consciousness? Or, on the other hand, as the late Susan Sontag professed in her text *Regarding the Pain of Others*, images of the body in pain have replaced pornography as the new taboo.[62]

— Yet still, one cannot escape the expression of pathos in Max Aguilera-Hellweg's *The Sacred Heart*.[63] In *The Sacred Heart* *[fig. 17]* he tells us the following:

Organ donors must die from brain death. In the final moment of life, if brain death occurs inside a hospital and loss of blood supply to the heart can be prevented, the organs can be kept alive by keeping the donor incubated and on a respirator. Timing is critical. A human heart is good for only four to six hours, a liver for eight to twenty-four. A kidney can be kept viable for as long as three full days.

— *Brain death certification is serious business and cannot be easily explained. It is difficult for the family of the potential donor to comprehend that their loved one is dead when they see the person breathing on a respirator. Specifically, it is hard to explain the difference between brain death and coma. Because of the incongruities, the sensitivities, the subtleties of this distinction, some states require certification of brain deaths by two separate physicians.*

— *The last document of the thirty-seven-year-old woman's medical record was a checklist of body parts, the inventory of organs to be donated for transplant or research, marked with an X, and signed at the bottom of the page by her husband, who had given his consent. I tried to imagine what kind of pain, what was the nature of the generosity he felt as he signed his name. But nothing could have been stranger than seeing the surgeon open up her chest, her heart still beating, her lungs taking oxygen, yet knowing she was dead.*

— With the same intensity of purpose, Vladimir Mironov's experimental research into bioprinting is a pioneering foray into the problematics of regenerative medicine. Like the stuff of science fiction, images of architecturally rendered vascular systems are ripe icons for twenty-first-century medicine and more *[fig. 18]*. Add to the mix his conversions of Epson printer hardware, which substitute ink with liquid cellular solution, and what results is a thinly "printed" tissue.

— What social values and aesthetic issues are embedded in these images, somatically or conceptually?

MAX AGUILERA-HELLWEG

— *I Don't Bite [fig. 19].*

CATHERINE WALDBY

— In response to Aguilera-Hellweg's *I Don't Bite* post and to Suzanne's regarding body anomalies, I think part of the reason these kinds of images are so difficult to witness is that they attest to the highly contingent nature of what we consider the human form. Part of what is so fascinating about embryology and embryo images is the quite inhuman appearance the embryo has for such a long period of its gestation and the fact that it looks at its earliest stages like so many other animals—so it recapitulates the finely balanced, potentially quite different evolutionary process that produced human morphology. When embryogenesis goes "wrong," deviating from the norm, it is a too-graphic reminder of the fact that we are all just one step away from monstrosity.

[public blog]

WENDY WEBB

— I would like to respond to Catherine Waldby's comments on Aguilera-Hellweg's image titled *I Don't Bite* and to Suzanne's posting regarding body anomalies. I think it is apropos to mention Julia Kristeva's theory of the abject. Though the abject is rooted in psychoanalysis, its application to cultural, media arts, and visual studies has shed light on our relationship to images that repel and compel. In Carol Clover's writings on gender in the modern horror film, she writes of our desire to experience/seek out/create/appropriate visual images that are repugnant, gratuitous, and thrilling. It is this "thrill" and shrill aspect that is relevant to scientific/bio-art imagery. Waldby comments, "I think part of the reason these kinds of images are so difficult to witness is that they attest to the highly contingent nature of what we consider the human form." Julia Kristeva's discussion of the female body as a cultural symbol of contempt (per memory from Barbara Creed's *The Monstrous-Feminie: Film, Feminism, Psychoanalysis.* Routledge, 1993), as it "swells, bleeds, lactates, and changes shapes" speaks to our unpredictable and dependent relationship with nature, which problematizes our mirage of order and "norm." The abject describes a critical feature of the human. As Waldby puts it—"it is a too-graphic reminder of the fact that we are all just one step away from monstrosity." It seems this arena of abjection theory could be applied to the controversial images of embryology and human morphology.

MARVIN HEIFERMAN

— Hello all:

— I've been interested to read, in a number of posts, references to the difficulties the "public" has in responding to art-sci imagery. Robin Marantz Henig's post about the public's response to images accompanying her articles published in the *New York Times Magazine* and Max Aguilera-Hellweg's comments about the difficulty people have had regarding his (remarkable, I think) images raise an issue I hope will be addressed somewhere, sometime, in the course of this conference.

— Since different disciplines have very specific philosophies about how and why images should be made, used, interpreted, and even spoken about, I'm wondering what we mean when we speak about the audiences for art-sci images. Having curated a number of art exhibitions in the past that have explored artists' responses to genetic issues, it became clear to me, repeatedly, that while the shows were popular with scientists, artists interested in sciences, and visiting groups of students and academics, the larger art world's interest in engaging with artworks that are science based is limited, at best. Without a doubt, art-sci exhibitions raise fascinating and important issues (as Giovanni Frazzetto has suggested in his post) and generate interesting and important discussions among small groups of artists and scientists (as this conference confirms). But why does the kind of imagery we're talking about in this conference have a hard time reaching and/or communicating with broader audiences? What I'm curious to learn from participants in this symposium is who are the audiences they hope to reach and engage through their imagery, and to what effect?

ANDREW CARNIE

— I think there are potentially big audiences for artwork generated by science and art collaborations. There is a great deal of pressure on commercial galleries to show artists with track records of commercial success. The "product" rules in commercial spaces, and private gallerists use public spaces to promote and legitimize artists they represent. This, I believe, holds back art that is, in a sense, research led, radical, or is too big to sell, or is in someway difficult to appreciate. Little of my own work has been shown in art galleries. I have, however, shown in the Science Museum London *[figs. 20, 21, 22]*, the Design Museum Zurich, the Amnesty International Headquarters, London, the School of Tropical Medicine, London, the Natural History Museum, and Rotterdam, etc. These venues do provide large, appreciative, alternative audiences. The Wellcome Trust in London has done quite a lot to remedy this with its past showing spaces. The new drive for research in the arts in British universities, through the Arts and Humanities Research Council, has also allowed more challenging work to be produced beyond commercial pressures, but showing such work is still difficult.

[public blog]

ROGER MALINA

— I have been following with interest the discussion, as it is an area that our organization Leonardo/ISAST [International Sociaety for the Arts, Sciences, and Technology] has been very involved in.

— I would like to raise an issue about the culture context within which the visual culture of bioscience is viewed.

— Investment in bioscience is very unevenly distributed among different cultures in the world—and most of the discussion comes from people in countries that have large biotech industries, where there is a general level of scientific literary and for whom the benefits of bioscience are well articulated and supported. North America and Western Europe are generally "technophillic" cultures.

— Leonardo is involved in a network called YASMIN around the mediterranean region. Most of the countries in this region are not heavily invested in bioscience—are not generators of substantial new science and biotech. And the benefits of bioscience are very unevenly distributed.

— I myself work as a scientist (astrophysics), and visual culture is deeply tied at the joint to the science I do. We use visual expression to communicate our science to each other as scientists and to the larger public. I am technophillic in this context.

— Yet it is clear the society we are building using the best technoscience available is unsustainable—and it is not clear that our political systems can respond on the time scales necessary.

— As we develop the visual culture around the biosciences, how do we integrate different contextual perspectives and questionings? Existing networks have the natural tendency to reinforce existing or dominant visual culture. How does visual culture around the biosciences contribute to building metaphors, imaginaries of a sustainable culture on this planet? I think this is a problem.

ANDREW CARNIE

— Dear Roger Malina,

— Many thanks for your comments. They are very relevant and important.

— It is easy to get so involved in what is, in a sense, a position of privilege and not look at the wider issues of equality and fairness and current planetary concerns. I think this is often easy to do when you are an artist and find it difficult to make the work one is making at any one time, because of financial constraints, etc. Life is often very much hand to mouth for artists. Yet as you say, it is much more difficult for others in other regions to participate. My involvement in environmental issues has always been strong, maybe more so in the past when and before I was studying zoology, and when I was a lad spending lots of time out in the countryside.

— One possibility to begin to solve some of the ongoing issues is to set up courses within arts institutions which look at these dimensions. I would like to see a degree course in sustainable arts, where the principle was to make effective, say, visual art but with a minimal environmental impact. I think this would be effective, interesting, and challenging, especially if the subject mater were itself sustainability. As a topic itself, it is a subject I have not entertained, but might in the future.

— Much of my work is done within the context of a university, and certainly I think we should build networks of production and consumption free of some of the real world constraints within these academic institutions, for the arts and other disciplines. However, the trend has been to consider financial implications more and more within such places, reflecting the same old dominant cultures.

— I would like to think more on this topic you have introduced.

ROGER MALINA

— Andrew,

— Thanks for your response to my post re the concern about the larger context of building sustainable societies—but also the problem of lack of dialogue with people who live in cultures that are not strong producers of new biotechnoscience, and where the benefits of bioscience are very unevenly evident.

— Yes, I think that the transformation of our societies needed to address sustainability (and social justice) are so profound that art schools as well as science education must address these in the very structuring of the curriculum.

— A related comment is that the biosciences cover a vast range of specialties—but much of the art and biology discussion tends to focus on genetic engineering. Many artists, however, work in the context of ecology, the environmental sciences, and now, of course, climate change—these all include bioscience issues. Surely ecology is a key science for artistic intervention.

ANDREW CARNIE

— I agree with you; we are, in a sense, on some extraordinary adventure in science and art. What we see through science in terms of visuals is astonishing and opens up a new world; it certainly doesn't seem reductionist to me in any way. As an artist, I try to ask questions through my work and try to get the audience to reflect on the science, but looking at the work is not dependent on an understanding of the science. The experience is still to be appreciated whatever the background of the viewer.

— I appreciate what you say about the work that goes into image production by the scientists on their images. My work, it seems, just moves their work on in different and, I hope, at times dramatic ways. I hope it to be anchored by some conceptual idea that I am trying to convey. The time-based element of my work is good at drawing in audiences with the ability to change their perceptions of what they are looking at as the work progresses.

— This forum seems to have been a good chance to share ideas and images. I appreciate your comments about the works you mention. I make as many works that are not about science ideas as science ideas. This, I hope, rejuvenates the work each time.

LEONEL MOURA

— One of the reasons why our kind of art is not yet being shown in the art scene derives from the fact that there are very few people among curators, critics, and gallery owners with enough scientific background to understand the works. People don't like to promote what they don't comprehend.

It's human...Anyway my experience tells that the public reception is quite enthusiastic. Hence it's just a question of time.

KARL GRIMES

— As an artist, primarily an image maker, working with the animal and human body on a macro and micro level with different research institutions and museum collections over the past decade, my picturing practices have always been modes of questioning, testing, understanding, and ways of making sense of my life.

— I join you from a lighthouse in the west of Ireland as I wait (and wait) to image a stretch of sea in a specific set of conditions of wind direction, force, and temperature to match one recorded on March 7, 1887, by the then-chief lighthouse keeper. The intent of this picturing practice is a visualization and representation of archival evidence, a series of images for a large project of mine on the methodology of R.M. Barrington, the nineteenth-century Irish naturalist, who over a period of twenty years pioneered the first mapping of bird migration patterns around the Irish coastline. Barrington secured the willing participation of Irish lighthouse and lightship keepers, marshalling them into a far-flung body of researchers and observers. Over two decades, thousands of migrating birds of various species were collected following nocturnal and daytime fog collisions with the lighthouse lantern beams. He then classified, measured, and preserved each specimen. His analysis, *The Migration of Birds as Observed at Irish Lighthouses and Lightships*, was published in 1900.[64]

— In two days my digital camera and the outdoors will be replaced with an electron microscope, the following day by an MRI screen. For me, there is no major shift in picturing practice involved; each device enables and limits me in different ways from dimensionality (including time) and framing to optical precision. The major shift is in the questions I ask myself and in the range of emotions I experience on a deeply human level. It is this space that I try to rescue intact in my work, a refuge from the statis of the laboratory and its ordered paradigm. My "specimen portraits" may reflect little of their scientific genesis, replaced instead by implication and suggestion. But, the view from the lighthouse will still be recognizable, in parts!

JD TALASEK

— Although I'm not actively participating as a panelist, I'm going to exercise my director's prerogative and interject a comment.

— In addition to earlier references to theater and cinema (Djerassi's plays and Cuir's reference to *eXistenZ*), I would like to add a timely example of dance and performance art to the mix for consideration. While in New York launching the symposium with Suzanne, I attended a performance of the Forsythe Company's North American premier of *You Made Me a Monster* performed at the Baryshnikov Arts Center.[65] The performance combined dance, vocals, and the creation of a sonic environment with an installation of disturbingly beautiful skeletal models made out of cardboard cutouts. The models, which the audience was encouraged to build onto, were "read" by the dancers as a score for their improvisations. Text provided in the space suggested that the work chronicled a body besieged by cancer. The idea that a model of the body created by the spectator was then responsible for influencing the dancer (thus creating the "monster") is intriguing because it suggests the power of models to influence, correctly or incorrectly, our perceptions of self, body, and disease.

— Also, the physical experience resonated with Max's comments on the response to his photographs. The dancers' vocabulary was tortured, and they danced in close proximity to the audience, creating noticeable discomfort in the audience *[figs. 23, 24]*.

RAPHAEL CUIR

— Hi everyone, As Miriam van Rijsingen as already said, picturing practices are "core business" in art history. Could we imagine any art history without images? I believe it would become the most boring thing ever. As Leonardo da Vinci emphasized and as we all know, images can speak more than words, especially in the realm of sciences. But I do not think art historians and scientists share the same point of view regarding images. Art historians do not build up images; they interpret existing images in which they have a peculiar interest. Of course, works of art could be seen as the scientific object of the art historian, and sometimes we need scientific imagery of an artwork, like an X-ray, to find out about what lies hidden behind the surface.

— On the contrary, scientists create their images. It seems to me that they are nearer to artists in their use of pictures than they are to art historians. Scientists build an image of reality that is more or less faithful, exactly as artists do. But no need to say that most of the time they don't share the same goal. But sometimes they do.

— Here is a woodcut from Vesalius's *De humani corporis fabrica*, (1543) *[fig. 25]*. I think that this is an amazing image, of the kind that stops me and elicits my reflection. When we think about it, this image is really of the like we produce now, made up with a lot of Photoshop allowing us to combine whatever forms we like. But this combination of sculpture and flesh into a representation is nearer to the beginning of modern printing than to computer generated images. It is, as well, an image from the Renaissance rebirth of anatomy, in the book which is embodying it. It is for us a paradoxical image: mixing the mineral and the organic. Undoubtedly this image does not belong to what Martin Kemp has called "rhetorics of reality," in spite of Vesalius's claim for direct observation of the corpse. Nonetheless this image conveys a rhetoric which, both for the artist and the anatomist, underlines the primary importance of the antique models in the Renaissance, here the Torso Belvedere. By the same token, this image puts in the foreground the taking into account by science of a beauty canon. "The body displayed in a public dissection should be as well compounded as possible [...] so you may be able to compare other bodies to it as to a statue of Polyclitus." (Vesalius, *Fabrica*, 1543, V, 19, p. 548.) As Giovanni Frazetto pointed out, visions of science have become narratives embedded in cultural phenomena.

[public blog]

SEMEIOTICA

— "Could we imagine any art history without images? I believe it would become the most boring thing ever."

— For a blind person, an art history without images might be better. What about sound, taste, touch, smell...? All senses other than sight have been exploited in service to art and design. Why not MORE appreciation and preference for art histories without images?

RAPHAEL CUIR

— In response to Semeiotica... yes, I'm caught being sightcentric...if I may say so. The way I've been trained in art history makes it a discipline centered on sight, and I'm probably into this because sight is a sense I do privilege. Of course there is a lot happening in contemporary art that has to do with other senses...but still working on the model of exhibitions where you go and see the works...I know blind persons go to museums...how they enjoy it remains a mystery for me, and you incite me to find out. Art history for blind persons is, for me, a challenge, but there's one famous means to achieve it: ekphrasis, which then emphasizes the literary aspect of the discipline...

DEVENIR

— "...scientists create their images. It seems to me that they are nearer to artists in their use of pictures than they are to art historians."

— I think what scientists do is to interpret images

created by their experiments, as much as to create the images. Can't we say that Vesalius's work is, like Leonardo da Vinci's, between art and science (or technology)?

RAPHAEL CUIR

— In response to Devenir's question,

— Yes, I think we can definitely say that Vesalius's work is between art and science (not technology though). Vesalius was very well aware of the quality of the illustrations he had been seeking for and able to secure for his book, and we know for sure that he worked closely with the artists, following all steps of the making of images, including the printing of proof samples sent to the printer, from Venice to Basel. These are woodcuts, and they surpass by far the quality of many following engravings...

CARL DJERASSI

— I would like to comment on Cuir's remark. Be careful when you refer to "science" as if representation in science were homogeneous in contrast to other disciplines. It depends very much what scientific pictography we are discussing. Chemistry—notably organic chemistry, which can hardly be described purely in terms of empirical formulae but requires structural representations—is rich in complicated pictography. Much of that pictography bears no real relation to the actual shapes of the molecules as deduced from X-ray crystallography but is just used for visual convenience and ease of information transmittal. All you need to do is to open any page of a journal like the *Journal of Organic Chemistry* and examine the pictures. You will note that a specific molecule—say morphine—can be depicted in several different ways, some two- and some three-dimensional, which offhand look very different but represent the same molecule and are used interchangeably by chemists, like different dialects for the same word.

— But here is an interesting extension. I happen to be a steroid chemist (or better said, I was one), and a couple of years ago, the Austrian post office brought out a stamp of me, which in the upper right corner has two mirror-image representations of the steroid skeleton as envisaged by organic chemists. But then observe the large image of my face in the background. It is made out of thousands of such steroid structures. In other words, the chemist, who usually synthesizes steroids in the lab, suddenly has his image synthesized out of steroids *[fig. 26]*.

RAPHAEL CUIR

— Well, no doubt I was citing a rather general view. I'm not even supposed to use the word "science" for Vesalius's time. So thank you for your thoughtful comment. I very much like the chiasmus[66] figure of the synthesizer scientist being synthesized. It's very interesting that a specific molecule can be depicted in several different ways. Could we say that, in some cases, the scientist possesses as much freedom as an artist? Being able to choose between two- or-three dimensional representation, between one style and another?

CARL DJERASSI

— Let me comment on your last two sentences: "Could we say that, in some cases, the scientist possesses as much freedom as an artist? Being able to choose between two- or-three dimensional representation, between one style and another?"

— The answer is no, because there are really no limits to an artist's freedom and there are severe ones to an organic chemist. I specifically used organic chemistry because that is a discipline where pictography is almost indispensable, even for communication among specialists. It is impossible to do it in words.

— And while we do have some freedom, as I made plain in my first intervention, it is nevertheless severely restricted. When we depict chemical structures based on X-ray analyses, there is very little one is "permitted" to do (other than colors or movement) because one represents actual reality. But most chemical structures used by chemists are used as a form of handwriting, and the range almost goes from shorthand to calligraphy and from one- to three-dimensional representation.

VLADIMIR MIRONOV

— I would like to introduce my short, provocative, and in some aspects futuristic essay "On Art and Science: Bioprinting & Pygmalion's Dream?"[67]

— It is safe to say that art and science can and must cooperate for the mutual benefit and enhancement of each other. They are unseparated, integrated, and essential parts of never-ending creative, cultural, social, and technological progress and innovation. This process makes us human and allows us to further advance as the most sophisticated species on this planet. Bottom line: I think visual presentation unites art and science.

LEONEL MOURA

— When we talk about bio-inspired art, it does not mean necessarily that we are inspired by images produced by biologists, but that we understand natural processes and mechanisms and apply them to art production. If not, what is the difference between an electron microscope image of, let's say, a beautiful bacteria and a landscape? Should we paint it like an impressionist? Or isn't it more interesting, and artistic, to study the bacteria and use the acquired knowledge to generate art works?

[public blog]

NICHOLAS RUIZ III

— How is it that the "bio-inspired" differs from modern romanticists' infatuation with Nature?

LEONEL MOURA

— That is exactly my point. It does if it is biology and the understanding of the living mechanisms that inspire our art works. It does not if we just contemplate, illustrate, or reproduce images of biosciences, astonishing as they may be. For some of us, the relevant is not to build another "Cabinet d'Amateur," now filled with objects and pseudo-objects of science, but to use and deviate science for art production. Bio-inspired does not mean necessarily inspired in nature, but more properly, inspired in nature sciences. It makes a lot of difference.

INGEBORG REICHLE

— Greetings from Berlin to fellow panelists and visitors,

— My background is the field of art history and *"bildwissenschaft"* (image science). With my research group "The World as Image" (*Die Welt als Bild*) at the Academy of Sciences and Humanities in Berlin, I look at images both in science and in art and ask about the epistemic value of images in contrast to texts or diagrams, and try to analyze the logic of images in relationship to the logic of texts. Though I guess I have more to say about artists in labs and ("epistemic objects" and "biofacts") in the next session, I would like to give some short answers to questions posed intially:

— Although most art historians assume that their discipline is based on the analysis of original works of art by the educated eye of the art historian, the truth is, art history is based on knowledge production through images since its early days (J. J. Winkelmann, eighteenth century[68]), using techniques like engravings, photographs, and many other media. Art History was born in days when people had their doubts that texts and words alone could be adequate to give insights into art. In the end, photographic reproductions of artworks were, unlike the originals themselves, well suited to the generation of new classifications in schools, motifs, and national styles. The assignment of individual works to their place in the historical epoch, and the subdivision of an artist's work into its various phases were ways in which to catalogue images. These new picturing practices made it possible to convert aesthetic artifacts into "scientific objects" with the help of a "technical" device. By means of standardized reproductions, artworks of the most diverse origin were freed from time and place and, despite differences in materials and size, could be converted into objects suitable for comparison. With the introduction of portable devices and image libraries, a transformation took place. An otherwise unmanageable number of artworks could be converted into a collection of research objects that could then be held in the hand.

— New directions in research, such as those offered by neurobiology and contemporary consciousness studies, certainly provide greater insight into the working of the mind. Likewise molecular biology continues to provide us with a better understanding of the structure of the living world, but today it seems to be clear that even with the amazing insights these new worlds of scientific imaging offer us today, such images must, of course, be understood as historical snapshots, bearing within them their own historicity. The influence of these images upon our understanding of nature remains an issue of social discourse, because these new scientific explanations of the structures and processes of body and mind do, however, challenge our conception and understanding of what we call "human nature."

— Getting familiar with some imaging technologies in the field of the so-called life sciences, my conceptualizations of visual modeling were challenged because, along with this staging of the body as text or even the idea of life itself in textual form, the biosciences are, in practice, above all, a site of enormous image production. Inevitably the question arises as to how the discrepancy between the numerous metaphors from writing and language surrounding the life sciences can be explained in light of the extensive implementation of the most varied image technologies. The molecular representation of life as we know it today is a product of two technological waves of development. In the first, instruments such as the ultracentrifuge, electrophoresis, X-ray structural analysis, the electron microscope, and techniques for radioactive marking and chemical analysis of macromolecules made it possible to depict components of the cell as well as molecules and to determine their chemical, physical, and biological properties. The symbol of this epoch is the DNA double helix. In the second wave, biological technologies dealing with macromolecules, particularly nucleic acid, were developed. Whereas the first phase of the development of molecular biology was set in the realm of biophysical and chemical analysis, the second phase dealt in principle with the cell and the organism itself.

— Visualizations from the field of the natural sciences are never simply illustrations, but instead represent complex phenomena, which in their formulation are always bound by the conventions of representation and the reigning vocabulary style of their respective period or time. They touch upon arrangements as to the ways in which respective scientific context captures knowledge in an image and ascribes to it an epistemological meaning. Visualizations and models are without question significantly involved in the formation of knowledge and have always been an integral component of scientific efforts and legitimate heuristic means of forming theories. Although theories, however, attempt to explain concrete empirical relationships, models in the natural sciences deal much more with model-based assumptions and structural analogies. Theories can be viewed or understood as systems of evidence that attempt to adhere to assumptions about interrelationships based on strictly logical rules of reasoning and that have to stand up to empirical verification. Models, on the other hand, reflect much more in their structure the inner relationships of a problem area. Visual illustrations have always been used in the natural sciences to make visible scientific relationships, to visualize theories, or to graphically capture the results of scientific experiments.

— However, images and the media that transport them have their own logic and play an important role in terms of what and how we see and perceive things: scientific visualizations arise as part of a complex interplay of different agents. They must be produced as part of a labor-intensive process of production and negotiation and are to a great extent constructed artifacts and do not simply depict or form reality and/or the "object" of the respective investigation or experimental environment. Even photographic or other optical recording techniques do not simply record the phenomena of nature, but rather fix the state of prepared objects for the production of a visual record. Graphic representations, too, do not directly depict measured data, but rather are translated or converted into other media and visualized in diverse presentational forms that can be expressed using various representational conventions: in the form of curves, diagrams, or complex image rasters or other symbolic representations.

— At the end of my contribution I add some ideas and questions from the field of my current project:

— The World as Image

— The history of world pictures (*weltbilder*) is, above all, the history of their representability. The concept of a worldview (*weltanschauung*) already implies that visibility is a constitutive aspect of our experience of the world. In addition, the concept of a "picture" in the phrase "world picture" signals the unavoidable connection between worldviews and visual media. For this reason, the concept of a "picture of the world" should always refer to concrete visual models. The terms "world picture" and "worldview" do not constitute the two poles of an opposition between abstraction and concreteness.

To begin with, every worldview necessarily involves pictures; furthermore, world pictures represent the visual condensation of a worldview. Neither aspect precedes the other; instead they are inextricably entwined in a circular process. Two main theses can be derived from this fact: first, world pictures are not implementations of antecedent worldviews. Second, there are no worldviews without imagery. The relevance of the study of images to the "picture of the world" lies in the variety of the visual manifestations produced by a world picture. These range from historical models, such as cosmological world pictures, to the most recent visualizations in the life sciences. This team's research will be concentrated on paradigmatic studies. These will analyze the various manifestations of world pictures as they are conditioned by culture, social environment, and medium.

— On principle, investigations of the "world as a picture" include systematic and historical aspects. Our research is focused around four different questions:

— How can the circular interaction of world pictures and worldviews be analyzed by the study of pictures?

— What role do concrete visual media play for the function of world pictures in determining action and knowledge?

— What rules determine how a specific world picture can become part of a collective visual memory?

— How does the emergence of new techniques of producing, distributing, and experiencing pictures affect the modeling of new world pictures? How do these new techniques have a retroactive effect on the experience of world pictures handed down from the past?

[public blog]

ROGER MALINA

— Ingeborg,

— I think your foregrounding of the "epistemic object" is useful—and raises the question of whether the "experiment" has the same status in science as it does in art.

— Elie During has written some about this topic, in particular regarding the "role of the prototype"[69] and there was a conference on the topic in Manchester in 2005.

— The concept of experiment and prototype has migrated from scientific fields to general culture, but the visual object is not always viewed the same way by artists and scientists. There is much discussion in the visualization community about the process of rendering images in a way that they are communicable to a larger public.

— This is compounded by the way visualization of

"computer simulations" are acquiring similar status to visualizations of data in experiments—there can be conflation of the epistemic object.

— The Manchester conference summary was: "Experiment-Experimentalism: An International Interdisciplinary Conference on Science and the Avant-Garde," Manchester Museum, Manchester, March, 11-12, 2005

— Artistic practices are often described "experimental," but what exactly is the nature of relations between art and the notion of experimentation in science? The conference "Experiment-Experimentalism" aims to explore this question by bringing out rigorous, parodic, and poetic connections between experimentalism in the sciences and in the visual arts from the early twentieth century to this day. Bringing together historians, art historians, historians of science and technology, artists, and philosophers, this conference will seek to track the migration of such terms as "research" and "experiment" from science to cultural spheres, and to address different issues and themes such as the emphasis on process over results, as well as notions of chance and risk in the avant-garde and science.

EUGENE THACKER

— Hi all - I realize this phase of the discussion is ending today, but I wanted to chime in, and perhaps also to offer some questions to contribute to the transition into the next phase on bio-artists and the lab.

— I agree with Cathy's points about digital media and biological mutability. I think you can also see it in pop culture. Many sci-fi films (often reinventions of comic-book superheros/overmen) will feature digital images of DNA. One thinks of the X-Men and Spider-Man series, Hulk, etc. This can either happen within the story world (e.g., a lab with computers displaying DNA as part of a mad scientist's experiment), or as a formal part of the film itself (e.g., visuals in opening credits of a film). Now, of course, DNA has, as Dorothy Nelkin and others point out, become a "cultural icon," standing in for a great many things. But I think one message of CGI-DNA[70] in pop culture is exactly to reinforce the notion of the mutability—and programmability—of the biological domain. Flesh has source code. I don't know, sometimes I feel that pop culture does more "radical" things than any art practice (I like Cathy's notion of "biological fantasy" here)...But I wonder how all this changes (if at all) when we move into the tissue engineering lab, where (as I understand) computer modeling is increasingly being used (e.g., to model biodegradable scaffolds for seeding cells).

— And here I turn to Oron's points concerning representation. On the one hand there's the notion that there is no "picturing" in the traditional sense with bio-art. One works with the very "stuff" of life, as it were. But then there's also the notion that what paint is for the painter, DNA or cells are for the bio-artist. The latter position makes it possible to view even a cloned mammal in terms of representation (painting, sculpture, etc.), while the former position, it seems, posits a kind of materiality that works against representation. (Oron, arguably your early work with cell paintings is the latter, while the *Extra Ear* is the former?) I wonder if the art world analogue for this position is performance art (or indeed "performativity" in general).

— In this way, I basically agree with writer/curator Jens Hauser, who argues that bio-art must be understood in relation to post-1960 performance art (and he includes body art and actionism in this). That is, the issues it raises are less about picturing and more about the effect, temporality/process, materiality. The specter of "presence" haunts this discussion of bio-art. Perhaps it's no coincidence that bio-art and "new media art" are happening at the same time, both grappling with the problem of mediation and materiality (telepresence and biopresence)...

— Ah, I see I've a lot of reading to catch up on, so I'll stop here for now.

ANDREW SOLOMON

— I would add this from where my writing about art[71] and my writing about psychiatry[72] intersect: that while we may look at images as an inspiration for art and appreciate their visual beauty, we should also be aware that they are in many senses created and imagined, and that how they are created has an enormous effect on public policy. Brain scans (fMRI) are numbers, thousands of numbers, which a computer then constructs into a photo-type image. It's not a photo, and is not made through the exposure of any sensitive medium to light. Imaging is probably the most promising area in psychiatry today.

— When I am asked what lies ahead, my first answer is always that while new medications would be great, the real progress will occur with imaging. With better imaging, we can work out subtypes of depression and schizophrenia, both multifarious disease entities. We might be able to see who will respond to which medications or treatments. Helen Mayberg, working at Emory, has used imaging to locate a previously unrecognized brain area that is disregulated in severely depressed patients, and has initiated a new surgical procedure, based on the deep brain stimulation used in Parkinson's, in which an electrode is inserted into this area, and this seems to regulate other areas of the brain that we have long known to be implicated in depression. She has achieved astonishing results: at least sixty percent of her patients show significant remission. Given that she works only with people who have failed on medication and ECT [electroconvulsive therapy] and who have no further options and who have been classed as permanently disabled by mental illness, this is amazing. What's really amazing in relation to this symposium, however, is that it has all come out of images and is explained to patients and media through images, and those images are at once incredibly specific and wholly artificial.

— Part of the struggle of the mental illness rights movement has been the quest for parity, to get government bodies and insurance companies to provide the same coverage for mental illnesses that they would for "physical" illnesses. Talk about why this is fair and just was largely fruitless, and then came imaging. Someone cleverly thought to code areas of brain activation in warm colors and areas of non-activation in cool colors. A whole vocabulary emerged that included phrases such as "lights up" to describe an activated brain area. As soon as you show someone a picture of a depressed brain, with those sinister expanses of blue where there should be pink or orange, the population accepts that these are physical illnesses. So every popular magazine article about depression or bipolar illness or schizophrenia or borderline personality disorder now shows brain scans, and lobbyists in the mental health movement take scans along when they meet with legislators. A picture seems to be worth a hundred thousand words in this enterprise.

— There are dangers attached to the practice. We will ultimately be able to image criminal brains as well and see what is activated and inactivated. If seeing the brains of depressives has liberated them from culpability and entitled them to treatment, what effect will such imaging have in the context of justice? And what happens when we can image the brains of people and show that they are physiologically (because personality is all reflected in brain physiology and activity) obnoxious or overbearing or nasty? The images contain truth, but they perpetuate the idea that there is a separation between physical and psychic reality. Physicality is an alternative vocabulary for psychodynamic processes, which are always brain-based. The images serve a political purpose admirably, but they are still a fiction; your brain is grey, and not pink and orange and blue, and these beautiful pictures are designed manifestations of numerical data. Like all tools of politics, these richly imagined pictures are powerful, and can be used for good purposes or ill.

[public blog]

DOLORES HANGAN STEINMAN

— As scientists in the field of blood flow imaging and working on computer visualizations, we are confronted regularly with challenging responsibilities at the interface of art and science. Due to the medical profession's increasing interest in developing less invasive visualization techniques, as well as a growing familiarity among the general public with digital representations, simulating real-life radiographs or angiograms is becoming commonplace. We were able to provide such images due to the continuous development of the technology (sometimes we question the "competition" between scientific curiosity and technological progress). We try to supplement the sensory perception and use the computer in replacing the recorded image by a man-and-machine creation. This process of generating computer models is therefore an interpretation of reality through the creation of easily accessible images following the cultural trend in place.

— The difficult task, we find, is that of generating images that are at once accurate and clear, are following established conventions, and are aesthetically pleasing. We are struggling to make concepts understood and accepted by both medical experts and the public at large while keeping the compass straight and being continuously aware of the ways in which this process can be manipulated to alter the perception of reality.

SUSAN SQUIER

— I want to pick up on Andrew Solomon's wonderful discussion of the ways that colorful fMRIs condense and fictionalize numerical data. Andrew, you suggest that "physicality is an alternative vocabulary for psychodynamic processes, which are always brain-based." While I agree that these images can give us another way to represent and discuss brain-based processes, I worry that the vocabulary they offer is an insidiously limiting one. The images omit, or, to be more accurate, they black-box, the factor of time. If they don't actually do so in the way the images are made (they can be taken over a span of time, or at a number of different instances), they certainly do so in the way the images are usually shown to the public as part of a persuasive presentation.

— Seems to me that the rhetorical power of fMRIs also smuggles in a pressure to replace the slow, unseen intervention that is talk therapy with the instantaneous intervention of deep brain stimulation, in part precisely because it can be captured by the "snapshot" visualization of this new technology. In its turn to evidence-based treatment, psychotherapy is finding ways to use the power of fMRI imagery to express its own ability to produce change. Presumably that will give psychology some leverage to argue for longer-term interventions. If there's anyone in this conference who can chip in here, I'd love to hear from them.

— However, in the interim, seems to me we also need to mark the power such images have to shift not only the modes of psychiatric treatment, but the time frame in which medical treatments are expected (permitted) to occur. This is emphatically not to argue that talk therapy is the only or even a preferable option in cases of intractable depression, but it is to encourage awareness of the fact that such images may also have powerful secondary rhetorical effects, such as catalyzing our gradual reorientation from a temporal to a spatial model of biomedical treatment. Or, to put it in other words, they may give more impetus to the institutional medical tendency to cut down treatment time.

[public blog]

DOLORES HANGAN STEINMAN

— I apologize for barging in, but I wanted to introduce the point of view of the biomedical imager with a clinical background. One advantage of the computer-generated medical images is that they decipher the "biological cryptogram" and make the unseen seen. They do not compete with data collected by the clinician but supplement them. As a research group, we are interested in the relationship between blood flow dynamics and vascular disease and the ways to convert this complex phenomenon into a mathematical description (or model) using Computational Fluid Dynamics (CFD). The reason we study haemodynamics is the fact that plaques and aneurysms tend to develop at sites where haemodynamics are most complex and the success of vascular surgery depends on positively "sculpting" (manipulating) the haemodynamic environment.

— Another advantage of the computer-generated images is the "viewing" of an organ as a whole, thus presenting the organ and its functions in the context of its interactions with the rest of the organism, and providing a better understanding.

— It is also important to note that since the images are generated on the computer, there is no invasion of the body (neither of its function /non-invasive technique, nor of the privacy of the patient— see, for example, Mona Hatoum's video *Corps Etranger* and threat of invasion and violation experienced by the subject). Of course, one can argue that images become the person and simulations are yet another way of dehumanizing the patient.

— As scientists and educators, we are also aware of need for "truth" in these images and appreciate the difference between the scientific recording of data seen as documentary (and, as such, supposedly safe of any manipulations) versus the scientific computer-generated simulations (created from patient-collected data that have undergone two conversions: once from the individual patient to numerical data (binary code), and then from numerical data thus obtained into images).

— Hope this helps a bit with the argument in favor of medical imaging.

SUSAN SQUIER

— Thanks so much for the added perspective on medical imaging, which really needs very little argument in its favor, seems to me. Its capacities are stunning. And yet when you say, "They do not compete with data collected by the clinician but supplement them," I would suggest you are speaking of the goal and the best case scenario, but not always the reality. Consider two frames from Brian Fies's graphic novel, *Mom's Cancer* (Abrams Image, 2006): these images convey the gap between the complex understanding of medical imaging available to the biomedical imager and the flattened reception in the medical encounter *[figs. 27, 28]*. And as for the use of images to dehumanize the patient, Fies's panels say it all in the ironic gap between the verbal message and the counter-intuitive, ungraspable meaning of the math behind the visual image.

— I'd love to see a discussion of the role of graphic fiction in talking back to medical imaging, in fact: one set of visual images for another. Take the comic panels, with their complex grammar and vocabulary (see Scott McCloud's works) and the medical images (with the rich grammar and syntax you discuss in your posting). It would be great to have a conversation between the two sorts of imagers: graphic novelists and biomedical imagers.

ANDREW SOLOMON

— Thanks to those who have responded here. I should emphasize that I think these images are very useful and that I am by no means opposed to them. They are useful in political contexts, as I've said; they are useful for patients who are trying to understand their own illness or that of their friends and relations; and they are useful for scientists because the logic of visual representation is important to various kinds of conceptualization. I am interested in the issues of artifice and of representation and so on, but I would not want to seem to disparage non-invasive imaging, which is one of the most important developments in medicine today. As I said in relation to Helen Mayberg's work, imaging is providing unanticipated solutions to what had seemed to be unsolvable problems. This is but one example of the remarkable progress that has been made. Still, the practices are misunderstood and literalized by the public, and it is important to examine the dynamics of those processes.

SORADA THUMRONGVIT

— I very much agree with you. Brian Fies's *Mom's Cancer* certainly fills "the gap between the complex understanding of medical imaging available to the biomedical imager and the flattened reception in the medical encounter." I grew up watching a French cartoon animation about the human body myself, and that was the first time I saw what a blood cell looks like. The conversation between graphic novelists and biomedical imagers you talked about might have already occurred there. However, I think it's only a matter of time before the medical imaging technology catches up and fills that gap itself; endoscopy, within certain limits, can produce quite a vivid image of the inside of the human body; ultrasound can now provide three-dimensional images of the womb. Even so, the graphic novel still has its own merits. The casual manner of it certainly moderates the issues that are usually hard to deal with conditions like cancer. On the other hand, finding a woman getting bald because of the chemotherapy in strip-comic format might be disturbing to some people as well.

— MFA Graduate Student SVA [School of Visual Art, New York]

SUZANNE ANKER

— At a time where the "self" has been characterized by philosopher Daniel Dennett as a "responding, responsible artifact," what role do images play in activating memory traces? All images conjure up undercurrents of emotion. Mirror neurons reflect the exterior world, somatically transforming it into interiority. Sometimes these somatically driven responses trigger feelings of empathy or fear, pleasure, or even panic. Although we know that what we may be viewing is a "critical fiction" (as in visual art or theater, somatic responses in many cases trump intellectual stasis).

— As processes of signification change with regard to the ways in which MRI technology creates pictures from numbers, I am interested in the ways that scientific iconography, defunct or otherwise, can be repurposed for aesthetic and cultural consideration. What intrigues me most about the early-twentieth-century psychological projective test, the Rorschach, is that not only has it become a cultural icon, but from a scientific prospective, the enormous extent to which these plates have been studied and employed.[73]

— By experimenting with 3-D software programs, I proceeded to see what would happen if a dimensional axis could be instrumentalized to revision the Rorschach inkblots into objects. Would they still appear as if randomly generated? Would they allude to possible new connections between the ways in which three dimensionality becomes a value-added strategy in both artistic practice and regenerative medicine (Vladimir Mironov)?

— What, in fact, resulted in my experiment were a series of sculptures that appeared to be very reminiscent of the natural world. The sculptures in some cases looked like animals and in other cases body parts and bones. Holding these sculptures in my hands generated within me an uncanny feeling of embodied imagination. That perhaps is my interest in sculptural form *[figs. 29, 30, 31]*.

JENS HAUSER

— Dear panelists,

— My background may explain partially my position: I'm actually working as a writer and art curator focusing on the interactions between art and technology, transgenre, and contextual aesthetics. My original profession before has all been about "imaging" for years. After film studies and studies in scientific journalism, I mainly worked for the European cultural TV ARTE since '92 and did documentaries and video. However, I trusted less and less how audiences can be intelligently "touched" by images. I moved to creative radio programs and projects to help access visually impaired people through audio-description. I have become more and more suspicious about the "epistemic power" of the image, probably because of what is referred to as the "pictural/iconic turn," and this skepticism is strongly influencing my curatorial work, leading to experiences with "wetwork."[74]

— Following this panel, it becomes evident that the majority of the participants seem to identify themselves with a cognitivist approach of how images can be interpreted (Goodman, Arnheim, etc., even if not mentioned explicitly), based on how images can be interpreted. I feel that the art I'm most interested in deals less with representation, but with staging presence. This is especially true when it comes to art in the context of biotechnology and biomedicine, as well as to the biopolitical topics to be addressed with a certain degree of critical distance. I here expressively join the points made by Eugene Thacker and Oron Catts. I would argue that knowledge production, if ever this is to be defined as a prime role of art, probably can be obtained directly or indirectly. Directly would mean through the epistemic comparability of visual art and scientific visualization techniques where relationships are formally established in the hermeneutic perspective of making sense. Indirectly would mean through a situation or encounter with art that is based on a phenomenological oscillation between meaning effects and presence effects. For example, art that engages the viewer/participant strongly enough encourages him to build epistemic, semiotic, and emotional clusters out of the experience.

— We may ask how art defines its aesthetic displays as metaphorical signs that refer visually to phenomena in the world, and more importantly, whether artists here even want to make rival use of the epistemological power of the image, or whether they see their role instead in the subversive questioning of dominant technoscientific concepts and dogma and therefore of their visualization strategies (as well).

— I feel that there is a strong need to produce effects of presence in our oversemanticized culture—which has independently been discussed by Hans Ulrich Gumbrecht in his critique of the central position of interpretation and his hypothesis that "any form of communication implies such a production of presence, that any form of communication, through its material elements, will 'touch' the bodies of the persons who are communicating in specific and varying ways." He underlines the fact that this has been "bracketed by Western theory building ever since the Cartesian cogito made the ontology of human existence depend exclusively on the movements of the human mind."[75] Therefore, I strongly feel that the art I want to promote is transgressing the semiotic procedures of representation and metaphor, and produces presence that cannot simply be mediated without reducing it to a purely heuristic placeholder of discourses.

— Here, the question of the audiences pops up. Leonel Moura and Andrew Carnie argued that the fact that sci-art is often shown in science museums and suffering from the predominance of commercial art structures is due to the lack of science literary of curators and critics that cannot comprehend it yet. Another way to see it would be that work in this area often is reduced to its inherent signs and may not oscillate between meaning effects and presence effects.

— In an age in which the techno-sciences themselves have become potent producers of aesthetic images,[76] the use of biotechnological processes as means of expression in art dealing with biotechnology may not have a prime depictive function. In my opinion, art that deals with biotechnology as a means of expression, as an example, distances itself from the assumption made on the comparability of scientific and artistic images as proclaimed in the context of the "iconic turn." It defies referential representativeness and illustrative simulation, such as that which characterizes the understanding of biological systems in digital media art. What seems to point to this "de-image-ing," which should also be regarded as often subversive, is the ironic approach that wetwork-centered artists may have towards visualization technologies such as GFP [green fluorescent protein] localization or gel-electrophoresis. When living systems are involved, or organic materials manipulated, are these only "living images"?

LEONEL MOURA

— I agree with Jens Hauser that many art-sci works are too complex and need some kind of explanatory complement, losing in the process the ability to communicate with large audiences. But this happens also with contemporary art that is not science based. I would even say that some art broadcasts incomprehension as a way to gain importance. The less the viewer understands, the better the artwork must be. And this was never a problem, on the contrary, for the art scene.

— I would consider another point, which is the fact that art-sci emerged from media art schools, labs, and universities and not so much from conventional art schools, bohemia, and art communities. The tough communication between the two worlds is not so much a question of code but more of object.

— Finally, I also agree with Jens on the question of presence. That is why, from the beginning and even more today, I am interested in bringing algorithms to the real world and not just to stare at the computer monitor. I have chosen conventional painting as a strategy. But the fact that autonomous and smart robots create these true paintings, in the sense that they exist physically and we can witness the painting's construction, generates a critical questioning about art, intelligence, and the human *[fig. 32]*.[77]

SUZANNE ANKER

— Michael Sappol has stated that medical iconography has a

"cumulative effect" on perception. He talks about the ways in which seeing certain images makes them more naturalized. What does "naturalization" mean in relation to picturing the embryo and the fetus? Brad Smith's embryonic visions dig deep into the recesses of evolutionary development as he exposes origins of body patterns cum totemic markers. At the same time, Catherine Walby, Carl Djerrasi, Robin Marantz Henig, Susan Squier, and yours truly have also invoked the image of the embryo/fetus in our work. I have stated elsewhere that the fetus is a primal marvel. Its representation continues to harbor political dimensions within the United States. Why have we not become "naturalized" to its presence?

BRAD SMITH

— First, I would suggest that we are becoming "naturalized" to fetal and embryonic images and representations. In Western culture, "Baby's First Picture" in the family photo album is often an ultrasound. Consider the moral, social, and political implications of this time-shift for having a new family member. One hundred and fifty years ago the new family member arrived at birth. Now the "arrival" is celebrated and ritualized at eight weeks postconception.

— I tread on delicate ground here because I am a man, although I was involved, if not more distantly, as a participant in bringing three children into the world. I wonder if the very concept of embryo and fetus 150 years ago wasn't very different from our concept today because of biomedical technologies in general and medical imaging in particular. Was the embryo or fetus in the more distant past a nonvisual experience, a visceral experience? How was the developing embryo or fetus experienced and interpreted by most of humanity when it was only felt, anticipated, and observed by and through the changes in the mother-to-be?

— Note that the vast majority of embryo and fetal images "erase" the mother and elevate the status of the embryo or fetus. These images emphasize the individuality of the new life and diminish the connection between the mother and embryo or fetus. My own research in imaging embryos is a perfect example of this consequential omission *[fig. 33]*.[78]

— Monica Casper in her book *The Making of the Unborn Patient* makes some excellent observations about how medical technology (and imaging) have altered our understandings, interpretations, and behaviors towards the embryo and fetus by giving it the status of "patient." She says, "In fetal medicine, the ultrasound image does not 'show' the mother but rather symbolically and visually excerpts and isolates the fetus from her body. In the domain of clinical decision making, these visual images of fetal bodies replace the material fetal beings still inside their mother's bodies in another part of the hospital or at home. The ultrasound representation is the phenomenon."[79]

— Embryos and fetuses now have their own doctors, their own lawyers, their own culture. I suspect that much of this social and political change is a consequence of suddenly "seeing" something that looks like an individual rather than perceiving this new entity by visceral means and as a gradual emergence of a new being through an ambiguous process.

— There are many other examples of how Western culture is naturalizing these images, including the Madison Avenue marketing campaign for automobiles showing an ultrasound and the tag line: "Is something inside telling you to buy a Volvo?" or the depictions of embryos and fetuses, whole and dismembered, by political groups arguing for or against abortion rights, stem cell research, and human cloning, or *Time [fig. 34]* and *Newsweek* magazine *[fig. 35]* cover photos to lure subscribers and purchasers.

— The embryo/fetus is now variously considered in our culture to be property, family member, medical cure, bearer of legal rights, experimental material, a potential, and an actualized individual. This new status of the embryo and fetus certainly correlates with the advent of new biomedical reproductive technologies and imaging, and might be a consequence of these technologies.

— Thus, I would suggest that we are becoming "naturalized" to the presence of embryo and fetal images.

KARL GRIMES

— In 1997, I first exhibited my project, *Still Life*, a series of large color photographic portraits of late fetal and neonatal bodies with a range of congenital abnormalities, at the MacLaurin Gallery, Scotland; the Gallery of Photography, Dublin; and later at the Art Exchange in New York *[figs. 36, 37, 38]*. My interest was to look at (and show) nature's asymmetry and question my own disquiet and fascination with these tender beings, bringing them to light and into the light from the historic medical collection and research laboratory.

In each country the reaction to the images followed a similar path. The gallery in Scotland was closed following numerous complaints and incendiary media coverage, and finally reopened following the intervention of the Scottish Arts Council, for viewing by adults only with the caution: "Viewers may find these photographs deeply disturbing." Dublin's Gallery of Photography blacked out their large public windows and insisted that visitors read a descriptive outline of the show before entering the galleries, also cautioning that the work may be"upsetting" and "not suitable for children." National tabloids followed with headlines such as, "How Low Can You Sink" and "The Grimes Reeper," etc.

The experience for me was not pleasant, but it did enable me to map the boundaries and survey heterosexuality's fortifications surrounding this "primal marvel," to quote Suzanne Anker, firsthand.

— Over the course of this project I spoke to hundreds of gallery visitors and different representative groups about the work's content. Their positions ranged from supportive to accusatory. The majority of responses were in the latter category, including many scientists, who viewed the work as an exploitative and violent infringement on the sanctity of our being (interestingly, other animals not considered) and/or the sanctity of the institutions of science and art. It was, in their view, pro-abortion, anti-abortion, an insult to women, to fatherhood, to beauty, to hope, to people with disabilities, to pregnant women, and deeply suspect on an ethical level.

— The sight of strange corporeal figurations inspires contradictory responses: recognition and denial, fascination and fear, identification and rejection, contagion. When a form departs from our preferences and desired fantasies, we gaze with unrest and suspicion. Comforting distinctions between what is and is not human become confused. Catherine Waldby in her recent post reminds us that "we are all just one step away from monstrosity." I agree.

— Margrit Shildrick's *Vulnerable Bodies and Ontological Contamination* offers a further insight on the anomalous human body and the subjects in Still Life, "The encounter with the others who define our own boundaries of normality must inevitably disturb, for they are both irreducibly strange and disconcertingly familiar, both opaque and reflective. They enable us to recognize ourselves; they are our own abject. As Grimes himself notes, 'Images of what we have denied turn towards us.' And once the initial shock of confronting what is usually excluded had passed, I found myself not repulsed, but moved to tears by the unaccountable beauty of the bodies. Beyond the marks of a violent and violating science that were evident in the confinement, both materially to specimen jars, and discursively to the category of abnormality, it was possible to acknowledge a siblingship which claims us." [80]

— In the end, it was and is just our mutated body on display in a culture where engagement with pathology and our bodies is highly problematic, save where that visual engagement is imbued with narratives of success and maps of past and impending conquest. Images from the biosciences feed well into this; they do not confront; they are "other"; we see a precision bombing screen effect, not our recognizable selves. We invite a good science story; we like it to deliver in dayglo. Our inner-colonial travels from micro to nano continents demand knockout home movies and travel snaps to communicate to diverse audiences and potential funders. Pain and mortality are distanced, triumphed in the glowing assurance of progress.[81]

EUGENE THACKER

— Now, as I understand it, "naturalization" here means "getting-used-to" or "becoming-accustomed-to" something that may at first seem alien or strange. So, for instance, a medical student in a gross anatomy lab must become accustomed to not only the dissected cadaver, but the correlated images of CT/MRI scans and—nowadays—D models of anatomy on a computer. Similarly, I, as a non-doctor, may see so many popular representations of CT/MRI images in TV shows like Medical Investigation or House that they may simply come to stand in for "high-tech tool" to me as a casual viewer.

— But how is this kind of naturalization of images related to ideas of "nature" itself? The interesting thing about the discussion thus far is that we're asking about the "naturalization of nature" via these images (e.g., of the fetus or embryo—and here I also think about the posthuman space-fetus in the last scene of Stanley Kubrick's *2001: A Space Odyssey*, as well as the references to oceanic wombspace in Andre Tarkovsky's *Solaris* - and coffee-table books like Nilsson's *Behold Man*). Does nature need to be naturalized? And if so, does this make it "ideological"?

— So this naturalization of nature entails a mode of training the viewer (specialist or not) to view in a certain way—or at least to become accustomed to the image so as to not question the act of viewing itself. Then the question is: what idea of "nature" and/or "life" must be assumed in order for someone to even be capable of being accustomed to such images?

— I'm interested in hearing from the visual artists here—what do you, as an artist, assume that the viewer will bring to the image in terms of preconceived ideas?

— I also wonder if, in addition to naturalization, we should be talking more directly about "aestheticization," or what it means to render biological life as aesthetic (as opposed to understanding biological life as sensory and phenomenal). Or if, in a sense, nature is precisely that which is aesthetic.

ANDREW CARNIE

— I had a spell before my dialogues with scientists of making work related to scientific imaging. I had spent a couple of years studying zoology before moving into the arts. Among the images I made was this work called Twins *[fig. 39]*. Its reference point for me was the embryo, or two embryos in fact. Twins never courted any negative response, though I think in many ways I thought it quite disquieting. The work is a photograph of two slices of bacon.

O R L A N

— Hi everybody!

— I'm sorry I'm late. I have been overwhelmed by the work for my retrospective and my book.

— New imaging technologies have always played a significant role in the conceptualization and the visual modeling of my work. I have always seized upon new technologies from every side. In France before Internet, we had the Minitel. I created a Minitel journal, Art Acces Revue, in which the goal was to open a space of creation where artists could take into account this new technology, interrogate it, in its lack, in its limits. Only seven tones of gray were available, and the definition was very low with big pixels. From Daniel Buren to Nam June Paik, from John Cage to Joseph Chiari, all invited artists created specific works, appropriating and decoding a space that was saturated by horoscopes, porn chats, cooking recipes, and the like.

— I made the first series of my digital self-hybridizations working with a graphic designer based in Montreal when I was in Paris. In France it was the beginning of the spreading of the Internet. The digital images were entirely conceived through e-mail exchanges. I would receive the sketch and send them back with my corrections and so on, until the image was matching my expectations *[figs. 40, 41, 42]*.

— I have always subdued the choice of my tools to my artistic project. Biotechnologies are part of the environment I put into question. An important reference was, for me, the text by Michel Serres "laicité" ["Secularism"], the preface to *Le Tiers - Instruit [The Troubadour of Knowledge]*, where the metaphor of the harlequin is used to defend multiculturalism and interdisciplinarity. I used this text for one of my surgery performances and in a series of reliquaries made of bulletproof glass, iron frames, and a few milligrams of my flesh *[fig. 43]*.

— During my residency at SymbioticA, the Art and Science Collaborative Research Laboratory at the School of Anatomy & Human Biology, University of Western Australia, I will create an installation (in progress) consisting of a large video projection (made of multiple diamond-shape images of living and dead cells from different origins), a transparent bioreactor especially designed, and a kind of harlequin coat made of plexiglas with petri dishes inscribed in diamonds. Some of the petri dishes will contain skin culture samples hybridizing my own skin cells and those of other humans and animals (fixed or in vivo). This experience will inscribe in the real and the living, the self-hybridizations I have undertaken with digital images *[figs. 45, 46, 47]*.[82]

— I have always enjoyed working on the fake and the real, the living and the artificial. I always think and work with the

"AND" instead of the "OR" dominating our Judeo-Christian culture, which is trying to introduce a type of complexity.

— Michel Serres, *Le tiers-instruit*, 1994: *The current tattooed monster, ambidextrous, hermaphroditic and mulatto, what can it make us see, now, under its skin? Yes, blood and flesh. Science speaks of organs, functions, of cells and molecules, only to admit at last that it's high time we stopped speaking of life in laboratories; but science never mentions the flesh, which, quite rightly, signifies the conflation, here and now, in a specific site of the body, of muscles and blood, skin and hair, bones, nerves and diverse functions, that inextricably binds that which pertinent knowledge analyzes.*

— *[...]*

— *Now already for a while many spectators will have left the auditorium, tired out by ineffectual theatrical effects, irritated at the turn from comedy to tragedy, having come to laugh and deceived at having been made to think; there will be some even—knowledgable specialists no doubt—who will have understood on their own terms, that each portion of their knowledge resembles the coat of the Harlequin, since each one works at the intersection of many other sciences and at the interference point of almost all of them. Thus their academy—its encyclopedic institution—formally rejoins the comedy of art.* [83]

MARCH 9–10

SESSION 2

ARTISTS IN THE LAB

— In Session Two: "Artists in the Labs," we will concentrate on the ways and means artists have turned their attention to working within the context of scientific laboratories and research institutions. Working with "wetware"[84] materials and processes, in addition to traditional and digital media, both hardware and software, artists are collaborating with scientists on particular projects. Other artists working independently also employ matter as their medium. SymbioticA, under the artistic direction of Oron Catts, is a unique research laboratory located in Perth, Australia, in which artists can experience firsthand the manner in which the biological sciences are being employed to realize projects in the arts.[85] Jill Scott's *Artists-in-Labs: Processes of Inquiry*[86] describes the first pioneering program in Switzerland addressing these issues as well. The nine participating laboratories in this project range from the fields of bioscience to physics to the computer and engineering fields.[87] In addition, there have been other projects sponsored by the Wellcome Trust[88] or even by individual artists themselves.

— In moving into this next area of discourse, please respond as you see fit. It is unnecessary to address each question separately or to answer them all:

1) What is a laboratory? What characterizes its distinctiveness, and how is that different (or not) from an artist's studio, a writer's den, or a scholar's office? Physically, these "sites of investigation" may be ensconced with technical apparatus, manuals and texts, picture archives, organic matter such as cells, plants, or animals, or inert materials and chemicals, tables and chairs; various energy sources, obsessive daydreams, and/or points of clarity which round out the décor. In some zones, the individual is alone; in others, collaborative teams work together on problem solving. As a locus of inquiry and investigation, what kinds of "things" are produced here? As behind-the-scenes places, to whom and under what conditions is public access offered?

2) What case histories of artists working in the lab can be cited as having seminal significance in developing new ways to conceive of art practice?

3) In another sense, laboratory practices bring into focus a host of other questions and obligations concerning:
3.1) animal models for and in research
3.2) experiments with transgenesis in animals and plants
3.4) reproductive technologies
3.5) biohazards and public health.

— These questions spring from the unprecedented and accelerating alterations to biological matter in all its guises. With regard to the social, economic, political and ensuing ethical issues brought about by these modifications, what responsibilities do artists and scientists have concerning the future of life and its variegated forms? To what extent and to what ends is hyperbole employed by both artists and scientists?

4) In a text edited by Jon Turney entitled *Science, not Art: Ten Scientists' Diaries*,[89] scientists record their daily activities and thought processes. What is noticeably clear is that scientists' efforts (similar to artists) in maintaining their practices are wrought with risk. Scientists in their journal entries appear as human as anyone else. Therefore, to the scientists out there, can you talk about the problematics of being a scientist?

FLORIAN DOMBOIS

— What is a laboratory? Maybe a place where "labor" (work) is carried out. In that sense we would have artistic labs as we have scientific ones.

— I just started a project, *Neuland*, at the University of Arts in Berne, Switzerland. In this project, historians curate examples of artists' work within a range of disciplines: visual arts, music, theater, design, and architecture. All of the artists in this collection have been or are actively engaged in "research."

— Many of the questions concerning animal models, transgenesis, reproductive technologies, and public health relate to the scientific way of investigating and depicting these topics. Therefore I think it is worthwhile as an artist not to follow the scientific tracks of depiction but to think and develop new explanation models to complete the scientific ones.

RICHARD TWINE

— Hi everyone,

— I'm reminded of the sociologist Mike Michael[90] describing the novel scientific construction of "bespoke" (as in custom-made) animals, which makes me think of the "newness" or otherwise of animal biotechnology. Clearly, in livestock breeding, notions of design (a humanist providence perhaps) are rather old, but technologies of genetic modification/biopharmaceuticals/cloning arguably introduce new possibilities of re/design, rationalization, and standardization not seen before, an approach to materiality that we perhaps see in other scientific domains as well, cosmetic surgery and nanotechnology for example. Is there a relationship, I wonder, between the movement of artists into the lab and this new bespoke value toward materiality in some areas of contemporary science?

— Is the role of the scientist changing? For example, do bespoke values encourage technical responses to political issues—I'm thinking of the cosmetic surgeon who helps (or tries to) the dysmorphic individual pass socially in terms of bodily (gendered and so on) performance, or the animal genomics scientist who constructs the possibility of a marker-assisted selection to produce less aggressive, more docile livestock animals.

— What happens when the artist enters the scientist's space and starts working with the same materials as the scientist? What is the relationship between artist intrusions here and those by anthropologists/social scientists/ethicists? Why are we there? Will we "go native"? I write as a sociologist who increasingly attends animal science conferences, reads animal science journal articles, and joins animal science academic associations. This is Bruno Latour's[91] notion of the hybrid academic for sure, but what are the consequences of this hybridity? Moreover, to what extent do we become mimetic of scientists? Are we then subject to similar ethical/regulatory concerns?

— Are we there to promote the social and ethical robustness of science, are we there to further our own discipline, and/or are we in a certain way of instrumental value to science projects? The "we" here is false, of course, as I don't wish to imply that artist and social scientists are necessarily entering laboratory space for the same reasons. But anyhow, hopefully some points worth raising.

JILL SCOTT

— To Richard Twine:

— I would like to react to some of your points as I have been thinking along similiar lines:

— "What happens when the artist enters the scientist's space and starts working with the same materials as the scientist?"

— Once, science critic Sandra Harding[92] defined the word "method" as a predetermined technique for the gathering of evidence or a set of materials used in order to carry out research, but the term "methodology" describes a theory and evaluation of choices about how research does or should proceed. Currently, these two categories cause debates among art and science researchers. These debates suggest that sharing methods might be easier than swapping methodologies and that learning in consortium teams leads to new discussions about these issues. While between computer science and media art a great deal of tool sharing is already taking place,[93] the different approaches to methodology often creates very different results. This is also the case for bio-artists who are working directly with materials from the lab. What happens when artists and scientists work on the same projects? Difficulties.

— "What is the relationship between artist intrusions here and those by anthropologists/social scientists/ethicists?"

— Many academically inclined art researchers are beginning to use ethnographical studies and workshops to analyze social questions and to combine the results into something called "proof of practice." This combination might also further legitimize the studies of the art researcher on a scientific level.

— For example, in the e-skin consortium, art researchers have learned to test participants' responses by using social science methodologies learned from the Department of Psychology in Basel.[94] Here science researchers are helping art researchers to empirically assess the navigation, information, and communication potentials of the users in relation to their particular levels of tactility, proprioception, hearing, and cognitive mapping. This testing can accurately identify inherent problems and inconsistencies in order to build the e-skin interface. In return, the media artist can benefit from the output of recording these tests and edit them into something digestible for the general public. Thus user tests about perception might empower the public with a deeper level of shared awareness, intimacy, and emotion. If one of the main aims in combining art research with the gathering of empirical knowledge is to "humanize science" for the general public, then perhaps these combined strategies are worth pursuing!!

— "Why are we there? Will we 'go native'?"

— Yes, probably, but it is an experiment that might yield interesting results, and research in both fields certainly need some thinking "out of the box." Perhaps we could be called hybrids. I know quite a few artists by now who love to gate-crash natural science conferences. Why? Well—it's inspiring.

— About your other questions, please see my posting response to Suzanne about our artists-in-labs project here in Switzerland.

CARL DJERASSI

— I would like to make two comments, both of them quite peripheral to the main thrust of this session:

1) The comment was made, "Scientists in their journal entries, appear as human as anyone else." In a way, that is a meaningless statement, since we are all human. So the question is one of definition. Take "journal articles." Scientists, more than anyone else, keep continuous "journals," which are their daily lab notebooks and they are also meticulously dated. They form a sort of diary, which in many respects appears "nonhuman" compared to the conventional idea of a "diary." Yet hidden among those lines and pictures are often burning personal issues.

2) The main point is that a scientist's behavior is quite idiosyncratic compared to that of other creative disciplines. Because of science's inherently vertical nature, scientists are both very collegial and at the same time often brutally competitive—frequently with the same colleagues with whom they are collaborating. I don't want to pursue this point (on which I

have elaborated in two novels[95]), since in many respects it is quite tangential to the issue at hand, but I think that careful definitions are warranted.

— But I would like to bring in once more what may appear a repetition from the earlier sessions, namely what activities such as playwriting can bring to the main theme of this second session.

— In my play *An Immaculate Misconception*, there is presented on the screen a simulated and yet real intracytoplasmic sperm injection. Why do I say real and yet simulated? In collaboration with two reproductive biologists, Drs. Roger Pedersen and Barry Behr, approximately thirty different intracytoplasmic sperm injections (ICSI) were filmed, each designed to produce a baby (with an ultimate success rate of ten to forty percent). But in collaboration with a film editor, I then spliced elements of different experiments together into the single ICSI experiment that was shown on the stage. Why? In order to dramatize them. For instance, in one of the thirty experiments, inadvertently, while trying to catch a sperm and aspirate it tail first into the capillary, the sperm swam headfirst spontaneously into the capillary and had to be ejected, its tail then crushed to make it immobile to then allow it to be aspirated tail first. There were humorous and serious metaphoric implications associated with the depiction of this incidence.

— In another experiment, at the very end, when the capillary has "penetrated" the egg and the sperm is about to be "ejaculated," the sperm essentially refused to exit the capillary and always hung in by the end of his tail. Only after three vigorous pushes did it finally end up inside the egg before the capillary could be withdrawn. Again the humor and the serious implication were not lost on the audience or theater reviewers, all of whom found this event the high point of the play. Yet the experiment was both real and concocted.

— A second example is the play *Oxygen* that I wrote with the distinguished chemist Roald Hoffmann.[96] It deals with the discovery of oxygen and the demonstration by the three putative inventors (Scheele, Priestley, Lavoisier) having to demonstrate their experiments before the King of Sweden in 1777 (an invented occasion to resemble a modern Nobel Prize). In a staging (now available commercially as an educational DVD) by the University of Wisconsin theater department, actual experiments were conducted by the actors under the tutelage of Professor Bassam Shakhashiri, one of America's great chemical demonstrators. In a German production, the director had the actors use life-sized puppets to conduct the experiments, which were semi-real in the chemical sense. In a Brazilian production, the experiments were not shown but only the results described by the wives of the scientists who were sitting on the stage commenting on the nature of their husbands' experiment.

— A third example is my play *Calculus*,[97] dealing with the priority struggle between Newton and Leibniz about invention of the calculus. In the play, I have the mathematician Abraham de Moivre, a colleague of Newton, explain differential calculus to a diplomat by having the latter time him while he ate an apple, to illustrate the rate of change at any given moment of a quantity that itself is changing in relation to another quantity.

— A fourth example is Stephen Poliakoff's play *Blinded by the Sun*,[98] where he illustrated the actual (mistaken) concept of "cold fusion," based on the misinterpretation of an experiment. This was actually based on one of the biggest "scandals" in chemistry during the 1990s.

— In summary, I realize that "Artists in the Laboratory" is supposed to discuss the opposite, but what about if the artist (i.e., playwright) also happens to be a scientist and brings his scientific laboratory into his artistic one?

— I can hear murmurs in cyberspace asking me to shut up, which I am now doing.

SUZANNE ANKER

— To Carl Djerassi:
Perhaps your reading of my comment concerning the public perception of scientists suffers from a too literal interpretation. Let me explain my statement more fully. Issues of public perception of both artists and scientists are relevant to our discussion.

— Depictions in films, literary works, and visual art have represented scientists' personas in less than favorable terms. Usually pictured as individuals seeped in hubris, the image of the "mad" scientist is still pervasive within popular culture and mass media. In these narratives, scientific experimentation usually goes awry and culminates in disastrous effects for communities at large. From Mary Shelley's *Frankenstein: Or the Modern Prometheus* (1816-1818) to H.G.Wells's *The Island of Dr. Moreau* (1896) to Aldous Huxley's *Brave New World* (1932) and *Ape and Essence* (1948), the early modern period is ripe with reproach for the righteous scientist. This attitude of mind of the sinless scientist who feels his work is "for the good of mankind" is further personified in more current work. From Philip K. Dick's *Do Androids Dream of Electric Sheep?* (1968) to Margaret Atwood's *The Handmaid's Tale* (1985) and *Oryx and Crake* (2003) to Steve Reich and Beryl Korot's opera *Three Tales* (2002), similar archetypical tropes are in play.

— Is this merely a coincidence, or is the public's perception of the scientist still a skeptical mix of fallen angel and rational soothsayer? Certainly the recent controversy concerning the legitimacy of Woo Suk Hwang's research with cloned human embryos creates a "public image problem" for the scientist.

With regard to scientists' diaries, I am not referring to process notes, laboratory records, or other procedural record keeping documenting the day-to-day working of an experiment. Instead, I am citing Jon Turney's (ed.) text *Science, Not Art: Ten Scientists' Diaries*, published by the Calouste Gulbenkian Foundation in 2003. Sian Ede, who unfortunately has not been able to join our interchange at this conference, wrote the foreword to this volume. In her absence, I quote her text thus:

The science world, as these diaries clearly show, is almost brutally competitive, and those on their way up are often obsessively occupied at their laboratory benches, computers, conferences, and field trips. Only when they have made it to the top [do they] have the time to be spokespeople for science in general. [In this collection of diaries,] we wanted to avoid the elder-statesmen stereotype and to find scientists from a younger generation who, though high-flyers, had perhaps emerged from less conventional backgrounds....The result is an extraordinarily frank and sometimes poignant collection of personal narratives.[99]

I cite one such scientific diary entry from ecologist and meteorologist Yadvinder Malhi, to clarify my point thus:

Thursday 17 January

In the morning we mark out old forest plots in a floodplain and dry-land forest, delighting in seeing armadillos, squirrel monkeys, giant millipedes, ground turkeys. In the afternoon I hike with Pedro, a likeable local Quechua guide, up and down trails to find a remote plot. The forest gradually soaks into my skull and I feel healthier and more centered than I have during the whole British winter. It's hard to explain. Something to do with being part of a world ancient and mysterious and feeling it in your skin and senses, not just in your intellect. I wish that Rachel, my fiancée, were here to share this.

Sunday 20 January

I go home tomorrow but the expedition continues for another five weeks. Despite exhaustion, the field team drags itself out for a farewell drink at a nearby snack bar. The bar turns out to be closed, but no one minds the effort. We laugh our way home beneath a crescent moon peeking out between the tall silhouettes of the forest trees, the air heavy with the fragrance of white ginger flowers.

Tuesday 22 January New York

After a week of forest isolation, I connect back into the global mind while changing planes here. The talk is of war: it rumbles on in Afghanistan and may move to Iraq; the Israel-Palestine conflict continues and there is the threat of nuclear war over Kashmir. Much of my wider family lives a few miles from the India-Pakistan border. The big world and its tragedies suddenly overwhelm me. Returning home means facing an enormous backlog of tasks. I have to submit a paper on "Carbon in the Biosphere and Atmosphere in the 21st Century" to a special issue of Philosophical Transactions of the Royal Society *by the end of next week. Then there are long-delayed revisions to another paper and two reviews to deal with, all of which should have been finalized months ago."*

RAPHAEL CUIR

— To elaborate on Suzanne's comment concerning depictions of scientists in films, it is striking to notice that the scientist is killed, or murdered, in movies such as *Metropolis* (1927), *Blade Runner* (1982), *Terminator* 2 (1991), or *I-Robot* (2004). In Metropolis, the scientist Rotwang is pictured as a kind of mad alchemist, whose laboratory looks like a strange gothic house in the midst of the futuristic metropolis *[fig. 48]*. His character is as dark as his project is supposed to be: creating life artificially. Historically, alchemists attempti ng to create the homunculus have long been persecuted by the Church, as they were seen as usurping God's power to create life. I wonder if this religious "blasphemy" is not still behind the film narrative. And both *Metropolis* and the *Terminator* series revolve around the theme of the savior, no need to say he is never the scientist

EUGENE THACKER

— Hi all -

— I was thinking about this relation between the artist's studio and scientist's lab. For some reason I was reminded of the string of Frankenstein creature-features produced by Universal pictures in the '30s and '40s. What still fascinates me about those films are the lab scenes—all bells-n-whistles, neon lights, and Tesla coils—what are all those bulky machines doing? (The one in the original James Whale version of *Frankenstein* is perhaps only outshadowed by the one in *Frankenstiein Meets the Wolf Man.*) If I saw that exact setup now in a gallery, I would probably like it quite a bit (especially if it was "interactive").

— It seems that the space of the laboratory is oftentimes regarded as a space of alienation for the nonspecialist. Can the same be said of the artist's studio? We have, similarly, a popular image of this (how many shoddy films of van Gogh vs. Gaugin have been made?), but one thing I see firsthand is how these two spaces—lab and studio—are changing in the context of work in fields like new media, HCI [Human-Computer Interaction],[100] games, etc.

— I know that the obvious question here is how the Latourian laboratory is replicated, transformed, or questioned in bio-art. I'd like to hear others on this.

— I'd also like to throw in the question not of the space of the lab or studio, but what it means to "dwell" in the lab or studio.

Martin Heidegger's essay ("Building Dwelling Thinking") is in many ways too mystical for me, but he does point to the consonance between building—dwelling—inhabiting: "Building as dwelling unfolds into the building that cultivates growing things and the building that erects buildings." What if we take this phrase "growing things" quite literally?[101]

— But this leads not just to what it means to build or to dwell or to inhabit, but how space is enscribed by those acts. There's a short passage in Marx's *Grundrisse*[102] where he talks about the difference between ants and humans. Both build, but only humans (he says) make models before actually building. Curiously, Marx doesn't connect this modeling to commodification. But in what way is the lab or studio a space of property, or better, a space of propriety? Labs are expensive, so is equipment, and so are ideas. So how do scientists and artists dwell in relation to propriety?

JILL SCOTT

— The artists-in-labs program originating in Switzerland[103] was established to give artists the experience of immersion inside the culture of scientific research in order to inspire their content and develop their interpretations. This allows the artists to have actual "hands-on" access in the lab itself, as well as attend relevant lectures and conferences. On the other hand, this program's mission was to help scientists gain some insight into the world of contemporary art, aesthetic development, and communication channels between science and the general public. Collaboration between both parties including an extension of discourse and an exchange of research practices and methodologies was encouraged. As art researchers who are running the program, we are learning, comparing, and shifting according to the results. What is central, indeed crucial, is that researchers in both art and science fields still retain a commitment towards the public and their subjects of study.

— We have had failures and successes in collaboration, but what is very obvious is that many artists may never have the opportunity to meet nor work alongside any scientists unless there is a growth in lab residency programs. We would also claim that we have fostered the development of new approaches to art and science research collaboration rather than art practice. If science educators are looking for more sensitive and poetic metaphors and communication skills, then art can help. For contemporary artists "real" information is not taboo, and they particularly like more socially conscious scientists.

— As the general public is mostly uninformed about scientific debates in many fields, we claim that transdisciplinary approaches may provide art researchers with solid raw materials, pertinent debates, and unique potentials in order to encourage critical analysis in the public realm and perhaps even affect social change. This requires that art researchers learn more about science so that they can produce more highly skilled, interpretive, and reflective artworks, ones that might not only gain more respect from science but also be more critical for future debates in the public realm. Scientific research has such a large impact on the future of humanity that it would seem irresponsible to not consider these potentials.

— The concept of "shock value" needs more discussion in relation to art and science. We think that an artwork has to be accurate about scientific content; otherwise the science community will not be engaged with the resultant work. One interpretational approach, which is very much rejected by science researchers, is the use of "shock value" (i.e., Stelarc[104]). In these cases, scientists see certain artists as uninformed and problematic, not only because they misrepresent their research, but also because they are reminiscent of tabloid-style journalism. This damages the image of scientific research. Instead, they prefer artists with more considered goals who are excited about the specific research being undertaken in the lab itself. Science also has it radicals or mavericks like Marvin Minsky[105] and Hans Moravec[106] or Eric Drexler,[107] who have reputations of creating problematic fiction to shock the public and illustrate their points. These scientists are not often taken seriously by our artists-in-labs research partners, as our scientists see themselves as standing in the middle of any informed debate.

— We claim that the artist has to first be exposed to the everyday activities of a particular scientific inquiry before they can interpret the results for the public. Historically, more informed interpretations have already had a valuable role to play if they were backed up by solid claims from the science community (Hans Haacke's work *Rhinewater Purification Plant*, which conducted gray water reclamation,[108] or Harrison's *Sustainable Food Source*[109]). It seems that informed interpretations can not only help to explore art as a catalyst but also improve public relations for scientists. Furthermore, art researchers' interpretations of ethical and social issues within scientific research may also help to generate a new level of discussion within the scientific community itself.

RICHARD WINGATE

— Jill, you mentioned the use of metaphor (visual and written) and the differences in their use in arts and science. I'd agree that there is a lack of training of science researchers in the use of a visual language, at least (maybe not quite the same thing). To me, this gives free rise to a fascinating absorption and recycling of stray elements of visual culture in visual scientific presentation. This contrasts with the written science research paper, which in its ideal form is a highly disciplined and sparse form of literature, entirely free of metaphor (well, that's the idea at least). Perhaps this is the origin of the ambiguity you've identified—we formally reject metaphor within primary written research but bask in the pleasure of a poorly balanced visual metaphor given the opportunity.

— I think that scientists would enjoy explicitly confronting the poetic motives behind what they produce. A comparative scientific and an artistic appreciation of an identical set of events might come up with parallel models that complement and define each other. I suspect that, in any case, we use this process of poetic/literal comparison in interpreting what we see—even if we couch these interpretations formally within a more austere language.

— Successful art-sci collaborations are indeed a relection of metaphor, but I'm not sure that a resultant artwork would ever be judged purely within a science arena. My utilitarian stance would be that the value of the exercise came within the process. I'm not sure which part/element of science is able to judge or afford respect to the product—an artwork that was useful to science might fail in so many other ways.

MIRIAM VAN RIJSINGEN

— Thinking about collaboration projects, I think there are already several (and very different) views on that in the postings, mostly when talking about the core issue of art-science practices. Giovanni Frazzetto mentioned the "immutable mobile" and authority changes, which is an inspiring notion I think. But Vladimir Mironov's concept of "unity" as well as the idea of the hybrid academic flattens the issue to a non-issue. And please explain hybrid. My own view is based more on "difference" and differences in discourses. I investigate the interactions of art and science, but also collaboration projects as processes of "framing." Dialogues or collaborations as well as art works that are produced through those practices re-frame, take the other practices, and review it with different "eyes," as it were. The viewer or audience of the work will become an edgy reader/viewer, forced to switch between frames.

— In relation to this, I have a question for Richard Wingate (and Andrew Carnie) about their collaboration. I liked the way in which you, Richard, described the importance of pictorial representation in/for your scientific practice: the nuances you formulated, the practice of sketching. How, would you say, are these nuances looked upon in your own scientific field; how much have you changed your disciplinary frame through the collaboration with Andrew? How, for example, is "beauty" discussed, if at all? At the same time, you stated that collaboration became possible through the "resonance associated with particular words"—through language. I have the idea that in Jill Scott's collaboration projects, language is also the first level of framing and re-framing.

— I have difficulty with the concept of "new ways." Apart from Suzanne Anker's investigations and perhaps Jens Hauser's, little research has been done (or published) about the embedded-ness of these particular practices in artistic traditions. One of our Ph.D. students, Danielle Hofmans, is working on that issue, and her first findings are that most works are much more heavily embedded in art historical traditions.

— Nancy Princenthal asked earlier for neuroscientific answers to the issue of the unconscious processes of perception and production. David Freedberg's response to that posting is very interesting, specifically as he is referring to "embodied simulation." I am not sure exactly what he means by that, but I am interested, because I think that we should look more to the issue of embodiment in perception and production. It is related to what is said earlier by some of the panelists about the issue and importance of "presence" and "staging." This is what I would call "performativity." These are not just issues that can be investigated in relationship to the perception of the "works" (that is fundamentally part of my own research), but also how performativity is working in the scientific laboratory as well as in the artist's studio (or the exhibition space for that matter).

— The work of Viennese artist Herwig Turk, for example *Blind Spot* or *Blind Date*,[110] in collaboration with scientists Günther Stöger and Paolo Pereira, are really investigations into these issues. His work reveals that the laboratory is a specifically staged practice, in which both the scientific objects and instruments as well as the researchers possess performative qualities. It is an embodied space. Also most of Jill Scott's artist-in-the-lab projects reveal the same qualities, as I watch the documentary of those projects on her DVD.

— In the line of this, I would like say that the kind of "things" that are produced in the laboratory are not only "epistemic things," as Hans-Jörg Reinberger[111] would say, but also bodies (see Herwig Turk's *Blind Date*) and subjectivities (as revealed in Jill Scott's projects).

SUZANNE ANKER

— "Epistemic things," a term coined by Hans Jorg-Reinberger,[112] has been mentioned by Miriam van Rijsingen and Ingeborg Reichle thus far. It is a term that many of our colleagues, however, may not be familiar with. This term has migrated from the natural sciences into the plastic arts and humanities as a novel way to think about discovery and flux. An "epistemic system," to my understanding, relies on the intrinsic properties and "power of material objects" as generators of possible outcomes in experimental practices. How can this term be applied to performativity and process in both the laboratory and studio?

MIRIAM VAN RIJSINGEN

— Dear all,

— I am happy to hear that the conference has been extended until the March 15, as I was not able to contribute much in the first four days due to an unexpected busy schedule.

— I want to join in again and think further about the epistemic value of the art practice, specifically when engaged with bioscience. I am also very interested in Marvin's and Nancy's thoughts concerning Suzanne's question. As for Suzanne's question about the "epistemic" applied to performativity and process in both lab and studio, I think this is a very important question indeed. I do not think many artists think about the epistemic a lot, or about their work having epistemic value. Jill does when she talks for example about e-skin consortium. But that's also a specific case of "go native" in my view.

— Hans Rheinberger's view on the scientific production of epistemic things is based on the idea that what experimental scientists do is "taking and making," that is "making available." He defines this further as related to Piece's sign system, but specifically understood as a fluctuating system, defined as a "continuous process" of signification.

— I could think of artists in bioscience as experimental researchers, who could define their practice as one of taking and making, and of making available, not in the sense of a definite object, but precisely as a practice of signification.

— I think also, for example, that the (beautiful, yet disturbing) images of Karl Grimes are a clear case of this process of taking and making, making available. I would call them perhaps instances of signification that, I think, generate knowledge, not only about the objects in the archives, history, etc., but also about ourselves, our psychologies, etc.

— More tomorrow. Have to think alot over.

EUGENE THACKER

— There's been alot of talk about epistemological issues and knowledge production so far. Let me ask a dumb question—why, exactly, are we assuming that epistemology supercedes ontology in the lab or the studio? Maybe someone can clarify this for me—in Rheinberger I don't get why the language of epistemology enters at all—it seems to be a "soft ontology," in which artifacts are not quite Heideggerian "things" but not simply abstract ideas either. I understand how the phrase "knowledge production" enters the picture, largely because of how scientists and artists are situated within a larger political economy of ideas (which in both cases are privatized). So in an example like population genomics—which we can really call biocolonialism— the issue isn't so much what counts as knowledge, but how the concept of an "essence of life" (a certain percentage of genetic variability and distribution) is linked to a political concept about foundational group identity (be it in terms of race, nation, or population— or all three).

— Among Suzanne's points, the notion that nonscientists are playing "catch up" is provocative—not untrue, but also not completely accurate. On one level, yes, this does seem to be the case in academia, but only in certain fields. Depending on how cynical one is, one could say that the biotech industry has inadvertently made possible a whole host of science studies books. A more attenuated view would say that there is a kind of intellectual ressentiment[113] in the way that academics respond, reply to, and react to events that are already happening. It seems that this is one problem of thinking about how bioethics can be truly effective.

— Now, I find learning about new scientific fields and technical innovation as sexy as the next person. But I also feel that, at the end of the day, there are often basic and quite familiar concepts that emerge from them. For example: RP technologies, [114] tissue engineering,[115] and "organ printing." Great stuff —better than science fiction. But it raises a question about the relation between "life" and "form." Is the key to engineering an effective, functioning, histocompatible organ in the computer-based model (*eidos*) or is it in the actualization of that model "in vitro" or "in vivo"? (These are arguably the two most valued forms of "zoe,"[116] or biological life, in biotech.) We might assume that for tissue enginering the proof lies in the actual biological organ—it functions or it doesn't. But it is the former (the model, algorithm, *eidos*) that is, in part, rendered valuable as a patent or as part of a technology. It seems that it is necessary to distinguish "form" from "life" while admitting their intermeshing (is this not the premise of Aristotle's *De Anima*?).[117]

— So let me also suggest that science and technology are at the same time playing "catch up" to issues raised by classical philosophy.

VLADIMIR MIRONOV

— Please, do not worry too much about the bioprinted organ's functionality. FDA and other state regulatory agencies will not allow doctors to implant bioprinted organs without proven safety and functionality. From another point, the assumption that living cells or tissues have no function is just an assumption. There is no scientific rationale behind this statement. I personally never saw such cells in my whole scientific life. The level of function, the level of maturation, and the level of cell and tissue differentiation can be variable, but we are working on this and we know how to do this. Moreover, we have clearly formulated, measurable, testable, and, I believe, achievable specifications.

— Yes, the bioprinted organs and tissues must be maximally authentic to natural human organs from the positions of both form and function. But sometimes perfectionism is an enemy of the good. I mean "good enough" is good enough. If it is safe, can help patients, and has a proven positive clinical effect then it is already okay. Synthetic implants and prostheses from ceramic, metal, or plastic or polymers are not perfect substitutes and not even close in functionality to natural human organs, but they improve and save millions of human lives. Bioprinting will allow us to escape undesirable scenarios when human organs will be gradually replaced by mechanical parts and devices, and humans will become more like robots than humans.

— From another point, xenotransplantation[118] from transgenic animals also is not the best option, because every organ contains resident adult stem cells that can migrate into the brain and potentially differentiate into nerve cells. Humans with porcine neural cells in the brain may be a very interesting biological experiment, but "chimerization" of human beings is definitely not the most desirable goal. Bioprinting of human organs from autologous human cells will allow humans not only to be healthy, but also will allow humans to continue be humans. This is definitely the most desirable goal.

JENS HAUSER

— My concerns deal with the dialectical construction of the interrelations of "epistemology/ontology," "representation/presence," and "lab context/lab (re)presentation." I find Eugene's following question, in reference to Rheinberger's notion of "epistemic things," really excellent and fruitful: "Why, exactly, are we assuming that epistemology supercedes ontology in the lab or the studio?"

— In the context of "wet" bio-art (which I'm most interested in), this question contains multiple layers:

— Does the "lab" context refer mainly to the way that artists gain knowledge in a laboratory set while considering their practice as "art as research," or do they need a laboratory set in order to display "biofacts"[119] in order to communicate with an audience?

— Is the display of a biofact more an effect of presence or of representation?

— Is the display of a biofact (and even the experience to produce it) meant to stage shiftings in ontological categories (semi-living, or else), or to the hermeneutic production of knowledge?

— A lot of what Jill Scott has been presenting refers almost to the experience of "artists-in-labs," embedded in a context enabling them to gain insight into scientific contexts as a strategy for "art as research." Only secondarily, this seems to concern the displays, or mediating instances; this insight nevertheless should be expressed in a form that audiences may engage with emotionally or cognitively. Therefore, we are not surprised that a lot of this "art as research" work is then presented as documentation (video, writings) of the research process, which means that the heuristic dimension to approach "epistemic things" becomes more important than the staging of ontological difference. Is the "nature" of art to be qualified through the learning period of the artist or through the concretization that can be experienced by an audience?

— I feel that this question indeed is all but new and can be traced back to previous reflections on "the blurring of life and art" (Allan Kaprow[120]), which ends up in Boris Groys's[121] argument in "Art in the Age of Biopolitics."For Groys, "art has shifted its interest away from the artwork and towards art documentation, as art becomes a life form and the artwork becomes non-art, but a mere documentation of this life form."[122] My problem here is that then the documentation/representation of the process of "art as research" becomes itself again a representational sign that only refers to "art as life itself," thus becoming the representation of epistemic things rather than their presence that can be experienced by those whom the artists are supposed to communicate with (an "audience"), because of their status as artists—and which in turn gives them access to the lab in an art/science context.

— Alternatively, the "lab" question pops up differently when we speak about the staging of a lab in the context of a gallery display such as *Disembodied Cuisine* by the Tissue Culture and Art Project (TC&A).[123] Unlike other installations focusing on the topic of biotechnologies, the artists here have hidden the lab away (though it was technically needed) under a black dome and emphasized the public's view of the semi-living entities rotating in bioreactors, and thus insisting on the ambiguous ontological status of those "biofacts." The lab condition itself here has not been visually placed in the center of attention in terms of simple theater props or as the represented context of knowledge production.

— Oron Catts and Ionat Zurr in their displays often use tissue culture techniques that in fact have existed for decades (Alexis Carrel).[124] This "ontological staging of stuff" could be seen as the representation of epistemic things and at the same time as the presence of biofacts. This interesting twist relies on the potential of TC&A's work, in my opinion. This may echo what Rheinberger himself picks up from Nelson Goodman when talking about the notion of "representation."[125] Transgenic art employing GFP may be read at a comparable level, as the GFP technique iself is very basic, while its staging of presence/ontology may strongly communicate with audiences.

— Another question in the context of artists-in-lab and "social and cultural implications of biosciences" is what Roger Malina[126] often mentions as the artist's capacity of "making better science." This notion is, of course, based on a primarily cognitivist and not on an aesthetic approach, and it has a dimension of utilitarianism which institutionally frames the making of meaning.

TROY DUSTER

— I would like to address the idea of imagining and then imaging the outcome of chimeric research as a barrier to permission to conduct such research. There is both a large body of fiction/literature and a plethora of pictorial images (art/film) of human and nonhuman fusions that go back for thousands of years. So what makes these last few years special is that we are no longer simply "imagining" centaurs and chimeras. Laboratories are producing chimeras. When review committees consider permitting or denying such research, they must imagine the possible outcomes.

— In the early months of 2006, biomedical researchers at Stanford asked approval from the Institutional Review Board in order to proceed with a research project in which they proposed to inject fetal mice with millions of human brain cells. The purpose was to study processes in the developmental brain that might shed light on Alzheimer's and Parkinson's diseases. The Institutional Review Board approved the research. The chair of the IRB committee later commented to the press – disclosing some of the reasoning behind the decision. He said, "We didn't see a problem – because once the mice started to act human, the scientists could kill the mice." This, it seems to me, would/could/should (and inexorably will) animate an artist's imagination about chimeric research and what it would mean to conjure up and represent some imagery of a mouse that was "acting human." What would this be? Envy or jealousy? Compassion or greed or altruism? What might be a behavior that constitutes the image of "acting human"? Attacking other mice when they accumulate more cheese than they need? Accepting that accumulation as the natural state of mouse evolution?

— No one knows. Carl Djerassi's reflections about and his experience of bridging the worlds of scientific and artistic production might make him especially well positioned to "imagine" such a representation. However, I suggest—depending upon that imagined image—future IRBs could be greatly influenced by it in their gatekeeping decision making. Building on a wing of Eugene Thacker's post, the image of the Frankenstein monster has long played a pivotal role in reducing the capacity of scientists to pursue certain kinds of human experimentation, long before IRBs surfaced in the 1970s. Movies like *Gattaca* and novels like *Brave New World* have played their roles in imagining and imaging a eugenic future. But now we have IRBs. What artistic representation will help shape their cognitive-emotive maps of futuristic chimeras? They will certainly be coming our way!!

EUGENE THACKER

— The example that Troy mentions is noteworthy—and the rationale even more so. I assume most on the list have heard about the so-called Vacanti mouse,[127] which, if you can believe it, has its own Wikipedia entry.[128]

— Oron, maybe you can add another angle on this, since, as I recall, you worked with [Joseph] Vacanti some time back? Personally I'd like to see a whole line of plush stuffed animals based on biotech—the Vacanti mouse, Dolly (which could be a play on the problems with cloning and mechanical reproduction—each stuffy would be a tiny bit different), even Eduardo Kac's GFP Bunny. Is this biosculpture?

— But Troy's example brings up science fiction again. Elsewhere, I've argued that SF is a kind of arbitrating discourse between the biotech industry and the general public; The Critical Art Ensemble[129] has referred to this as the "promissory rhetoric" of the biotech industry. Here I mean "science fiction" not so much as a set of narrative genre conventions, but as a kind of "imaging" practice that involves imaging the future or possible future scenarios. There is a naive wing that unflinchingly practices the utopian version (Monsanto promising unlimited free rice for all underdeveloped countries). But there is also the more critical—but still conservative—wing that uses the dystopia to both caution but also support the idea of a biotech industry (I would say that the Stanford IRB example might fall into this category). And then, of course, there is the good-old-fashioned struggle of the free human spirit against instrumentality (I would put *Gattaca* here). But in all cases there is a certain kind of "visioning" practice involved that, arguably, is the core of what takes place in SF.

— Finally, for those who still think that biology is unfashionable, consider the BBC news story about wedding rings being grown out of material removed from the jawbones of engaged couples.[130]

SUZANNE ANKER

— Great idea! When shall we start this project *[fig. 49]*?

EUGENE THACKER

— Ha, that's a great image Suzanne. As soon as the NIH ELSI program gives us a grant, we can begin our, um, research.

— Ah, consumer culture is great, isn't it? It can cause you to post messages that have nothing to do with the list whatsoever.

[public blog]

URI

— There are already many such plush toys, both of bacteria and other biological "playthings," for instance this giant flesh-eating bacteria: http://www.giantmicrobes.com/us/products/flesheating.html) or this site that specializes in the study of internal organs: http://weebs.org/weeberworld/index.htm.

SUZANNE ANKER

— Dear Uri,

Thank you for your informative posting. What is your personal interest in this matter? Flesh-eating bacteria, the ebola virus, and the avian flu, to mention a few, are certainly entities that require respect. As molecule bombs, their infiltration into human (and animal) flesh raises the question as to their function as popular culture motifs. Do these plushies, in fact, masquerade and redirect our fears about these mortal matters?

MIRIAM VAN RIJSINGEN

— I like the stuffed plush, too! But I cannot see the "true version of a nightmare" or evil eye in Cthulhuu.

VLADIMIR MIRONOV

— The Vacanti mouse (some people claim that it is the second-most famous mouse after Mickey Mouse) was done from biodegradable polymer seeded with cells, and it somehow (probably as result of post-transplantational polymer biodegradation and tissue remodeling and contraction) could not maintain ear shape after implantation in vivo. There is no successful clinical translation after ten years. We, together with our Chinese colleages from Tsinghua University, Beijing, China, are using nonbiodegradable, FDA- approved polyurethane, and the results are much better. We did not see any change in the ear shape during several months after implantation. Cosmetic effect is obvious. Is it an art (biosculpture) or not? It is up to professional artists to judge. If we can define art as something that could be placed in a museum, then it is art. I saw the transgenic obese mice in San Jose science museum.

ORON CATTS

— This is an interesting thread and I would like to get into it on both fronts—the Wiseman human-brained mouse and Vacanti's mouse. But first, I would like to add to Eugene's SF comment.

— As mentioned elsewhere, much of the developments in the life science and in particular in the area of molecular biology, are overhyped, sometimes referred to as DNA mania [Andre Pichot[131]] or Genohype [Neil Holtzman[132]]. The interesting thing is that, as a result, both the opponents and the proponents of developments of biotech subscribe to the same hyperbole rhetoric that exaggerate the power of this technology—hence we have a debate that creates unrealistic expectations, as well as fears that have very little to do with the actualities of the knowledge and its application. One of the best examples was the use of the ear-mouse in an ad in the *New York Times*,[133] where the mouse was used as an icon of the monstrosities of genetic engineering. The problem is that, as you know, this mouse had nothing to do with molecular/genetic intervention. How can we then have a credible debate?

— Ionat and I just finished a paper addressing it and in particular the ongoing struggle we have to distance our work from the discourse of GE [genetic engineering] and molecular biology.

— As for the ear mouse—Vladimir rightly mentioned the failure of Vacanti's mouse to produce the ear that would keep its structure. This poster boy of tissue engineering can be seen as one of the most celebrated technological (note—not scientific) failures. The question is, did Vacanti know at the time that it was not going to work? Did he release the image, aware of its highly evocative nature, to show the potential of tissue engineering and call attention to the new ways of dealing with living materials? I know that it provoked me and was one of the major influences on my decision to work with tissue engineering as my medium of artistic research. One of the reasons for that was that this mouse represented the surrealist project come to life, but by someone who did not call on the history of these types of images. In a sense, like any "good" "bio-art" piece, its strength was in its eventual failure - making it a non-utilitarian, culturally evocative object. In my book, that is as close as one can get to an art piece. I still have a dream of trying to collect artistic references to the ear mouse for a show.

— Vladimir talks about a transgenic mouse at the San Jose science museum (as well as in the science museum in Shanghai that had a mouse with an ear on its back). It was made specifically for them and was presented alive for awhile. It is now preserved and presented (interestingly enough) in the section of the museum that celebrates our "genetically engineered future."

— Irving Weissman's human-mouse hybrid was even more explicitly presented (at first at least) as a theoretical object for debate.[134] This project (among other things) prompted Ionat and myself to develop our new project (see project description below). But before that I would like to comment about Jens reference to epistemology vs. ontology. It is interesting to note that when we started to work in this area, we thought that we are dealing with an epistemological crisis; however, in the last couple of years, we realize more and more that we are dealing first and foremost with ontological questions. It is not \ surprising then that we see our work as dealing with life rather then cognitivistic approaches of art and science (back to knowledge production vs. meaning production).

— Here is the blurb about our new project: *NoArk is a research project exploring the taxonomical crisis that is presented by life forms created through biotechnology. NoArk will take form as an experimental vessel designed to maintain and grow a mass of living cells and tissues that originated from a number of different organisms. This vessel will serve as a surrogate body to the collection of living fragments, and will be a tangible as well as symbolic "craft" for observing and understanding a biology that combines the familiar with the other. As opposed to classical methodologies of collection, categorization, and display that are seen in natural history museums, contemporary biological research is focused around manipulation and hybridization, and rarely takes a public form.*

— *To create NoArk we will use cellular stock taken from tissue banks, laboratories, museums, and other collections. NoArk will contain a chimerical "blob" made out of modified living fragments of a number of different organisms, living in a techno-scientific body. In a sense, we will be making a unified collection of unclassifiable sub-organisms.*

— *NoArk will critically examine and make strange contemporary life sciences formations that confront our familiar ordering of the living world: How do taxonomical systems based on traditional classification accommodate life forms created by humans? We hope to answer this through the development of strategies to collect, display, and preserve sub-life (lab-made cellular life), visit and research natural history collections/practices, consult with experts in regard to a place for sub-organisms in current taxonomy, and research and collect what widely are referred to as anomalies that do not adhere to traditional classifications.*

— *In addition to the philosophical and ethical dialogues that we feel this project will engage with, we are also interested in artistic and technological strategies for maintaining and exhibiting living collections of sub-organisms in a vessel for long periods of time. NoArk will offer an evocative tangible "semi-living" system which will further problematize the human anthropocentric desire to classify the living world around it. NoArk will present ecology of parts as an attempt to observe the living world through a post-anthropocentric system; once we have become fragmented into tissues and cells, the possibility of being embedded within each other, and within the greater living world, emerge; the human is not the center but rather a part of a larger changing ecosystem.*

— *Ultimately, this will be presented as an installation in which the vessel containing living tissue constructs will be displayed as a chimera, alongside technologically preserved specimens of organisms. The project will involve developing methods of co-culturing different cell types from different organisms over 3-D matrixes inside a costume-designed vessel.*

— *NoArk will present materiality of life and its forms, in a time that life is becoming a raw material for utilitarian manipulation. We believe that artists should offer alternative modes of engagements with the material of life.*

— *Description:*

— *Rapid developments in the life sciences and its applied technologies have created new ways for beings to come into the world and new categories of existence that are challenging the order of the world. This requires us —humans—to rethink our understandings and our relationships with our own identity/ body, other animals, as well as the concept of life itself. The growing number of "labmade" life forms, either modified living organisms or different combinations of modified living fragments such as cell-lines and tissue (which we refer to as sub-life or sub-organisms), requires special attention.*

— *In pharmacological factories, research universities, and other technologically driven institutions, there already exists a mass of disassociated living cells and tissues (sub-life) in the thousands of tons. These fragments do not fall under current biological or cultural classifications. We created the Tissue Culture & Art Project (TC&A) in 1996 mainly as a way to define this category of life and, at the same time, as an attempt to destabilize some of the rooted perceptions of the classification of living beings. We see TC&A, and our other attempts to grow aspects of the extended body, as an amalgamation of the extended human phenotype—a disembodied body that is unified in living fragments, and an ontological device for reexamining current taxonomies and hierarchical perceptions of life. The extended body is by no means a fixed, scientifically binding order; it is, rather, a soft, artistic, and conceptual view of the subject of technologically mediated and augmented life. NoArk is an attempt to develop this idea further by engaging with the notion of the collection of parts that constitute a whole.*

— *By creating a device that will allow the co-culturing and fusion of cells and tissues from different genotypes and phenotypes (i.e., from different organisms and different tissue types), NoArk will present the breakdown of both Linnaean taxonomy and Molecular systematics (chemotaxonomy). A new technologically mediated ecology of semi-living fragments will question deep rooted perceptions of life and highlight the need for re-evaluation of human relationships with the greater living world.*

— *The new sites for the collection of specimens of "neo-organisms" are the life science/engineering laboratory, the research hospital, the biotech industry, and, increasingly, among artists and amateurs/hobbyists. These specimens of neo-organisms and sub-organisms are catalogued and collected systematically, in tissue banks, research institutes, and the patent office. However, most of these systems have little connection to historically agreed upon taxonomies of the natural world. The appearance of these new forms of life in the public arena is, say, more akin to the cabinets of curiosities then to the natural history museum, and it is almost always anthropocentric.*

— *As part of its historical narrative, NoArk will investigate the construction of knowledge through the acts of collection and classification as manifested by natural history museums, which stand in opposition to the disorganized and unique conglomerations found in cabinets of curiosity. We will also contrast these two historic attempts to systemize life to the development of modern biological curiosities, bringing into question deep rooted perceptions and beliefs about the ordering of life.*
In NoArk we will explore collections of natural history museums, tissue and cell banks, and biological laboratories.

— *Biology -Cell and Tissue Culture:*
NoArk will house cell-lines that we will obtain from various cell and tissue banks. Cell-lines can broadly be defined as modified cells (often immortalized) that are derived from primary culture (cells and tissues that are taken from complex organisms).
An established or immortalized cell-line has acquired the ability to proliferate indefinitely, either through random mutation or deliberate modification. There are numerous well established cell-lines representative of particular cell types, from a wide range of sources. We are particularly interested in cell-lines that are established from cells of two or more individuals, and cells in which their origin is designated as one type of organism while being classified as another (such as the McCoy cell-line that originated from a human and is now classified as a mouse cell-line). It is interesting to note that there was at least one attempt to classify a cell-line as a new type of organism that should fall under traditional classification, Helacyton gartleri *(Van Valen & Maiorana 1991).*

— *For the last ten years we have worked with, and in some cases developed, environments for cells and tissues to grow within. These environments are often called bioreactors, and we refer to them as the Techno-Scientific Body. For this project we will be interested in working with new types of bioreactors, as well as developing a prototype vessel/bioreactor that will allow us to dynamically co-culture different types of cells. The bioreactor will act as the visual foundation for the conceptual underpinning of the project while also functioning as a vessel for its enclosed cells.*

— *Being located at a lab in a biological science school within a research university will enable us to order the cells of interest without breaching any regulations and material transfer agreements (we feel that it is important to note this, in the light of artist Steve Kurtz's case, in which he is facing court for getting biological materials that were ordered on his behalf by a scientist).*

— *Having ten years experience in tissue culture and tissue engineering, we are now interested in perfecting our techniques for co-culturing and fusing cells from different species. We will analyze the successes and the types of fusion using the expertise of Dr. Stuart Hodgetts. This stage will be monitored and documented using the microscopy facilities available for us in the School of Anatomy and Human Biology, the University of Western Australia. We will also learn new techniques in the Departamento de Engenharia Biologica at the Universidade do Minho, Portugal. This part of the project will help us to scientifically, conceptually, and metaphorically learn more about cells interrelations inside a techno-scientific body. Growing the cells over and into different matrixes will be an integral and important part of the biological research as well as towards the final visual presentation.*

— *We will be developing the prototype for NoArk using both modified scientific equipment as well as of-the-shelves materials. We were fortunate to get the use of a bioreactor for free for a year. This is an expensive and significant piece of equipment that will enable us to learn about methods and technologies employed to grow cells externally to their host body. This bioreactor represents a novel way of growing tissue in a clear, soft environment; it is also unique as it is a stand-alone bench-top bioreactor that does not requires an incubator for its operation, and all of its operations can be controlled from a computer. Working with the Wave bioreactor will enable us to look at ways of modifying the system for the development of the first prototype of NoArk.*

— *Working with Dr. Clive McFarland from the Biomaterials and Tissue Engineering Graduate School of Biomedical Engineering, University of New South Wales, will enable us to research the use of hollow fibers as both delivery system and a substrate for the cell and tissue growth. His expertise will be invaluable in the development of the vessel and other aspects of the NoArk prototype.*

— *The development of the bioreactor represents another possible benefit to the development of low-cost biomedical equipment; we will be developing a cheap, large-scale bioreactor for co-culturing cells and tissue, and might be able to come up with some novel approaches that will be able to be used by other researchers and artists.*

— *Taxonomy:*

— *The exploration of strategies to collect, preserve, and display neo-life and sub-life will benefit both the emerging field of biological art and possibly will contribute to collecting and preserving specimens of lab-made life forms. This will allow for the development of systematic approaches to be adapted by research institutes and museums who wish to collect and preserve these types of specimens.*

— *This will be conducted by talking to specialists in the Western Australian Museum (as well as other museums around the world) and having accesses to methods as well as collections in the natural history museum. We have formed strong relationships with the Western Australia Museum, which holds a vast collection of animal specimens, including marsupials, platypuses, etc. These unique animals by themselves presented a taxonomical challenge. The WA Museum will assist us in developing strategies of collection and preservation using their well established procedures and protocols.*

— *We have found a few locations that have begun to tackle the issue of systematically collecting sub-life and its orderly preservations. Examples are ATTC (the American Tissue Type Collection) – a global bioresource center (which most laboratories around the world work with)—and the San Diego Zoo's Bioresource Banking project, and in particular the Frozen Zoo and Adult Stem Cell Acquisition and Culture programs that are positioned in a very interesting juncture of collection and classification. We will attempt to use the Bioresource Banking project as a benchmark to question its epistemology, and explore its rhetoric and strategies of collection and preserving sublife. None of these resources have explored artistic sub-life or artistically driven public displays of the new sub-lives and their peculiar position in the continuum of life.*

— *NoArk will be displayed as a semi-living artistic vessel alongside preserved living organisms (specimens) arranged to reflect new taxonomies. It will offer a visual interpretations of the so-called "new order" by presenting, maintaining, and growing our living and semi-living collections, as well as preserved ones. The exhibit will form an historical narrative of the evolving living world and the human position within this "chart." Special techniques will be devised, mainly by the use of hollow fibers to physically and conceptually "connect" between the life in the vessel and the rest of its surrounding specimens, to suggest alternative taxonomy that accommodates neo-life and sub-life.*

— *We also hope to be able to use sensors that will be able to provide with live feed information on the well being of the sub-life inside the vessel. This information will be then transferred to a dedicated website. We believe NoArk has the potential to be exhibited in art galleries and other public spaces, such as natural history museums, around the world as well as the new locations of the cabinets of curiosities: science fairs, zoos, performance places etc.*

SUZANNE ANKER

— Oron Catts, Artistic Director of SymbioticA, has rigorously worked on developing a unique laboratory/artistic residency and educational program for artists engaged in "wetware" practices. Oron, can you describe the ways in which artists, who may or may not have had experience in "wetware" media, begin to develop his/her project? What guidelines or teaching tools or workshops are offered to aid the laboratory novice reach his/her artistic goals. How does your residency program work?

ORON CATTS

— Dear people,

— As I told you, I am currently presenting one of our living installations, Victimless Leather, as part of the show Free Radicals at the Israeli Centre for Digital Art. Showing something living is somewhat different than visualizing it and requires quite a lot of time tending to it. That said, the tissue seems to grow well, and I now have time to finally focus on this very interesting exchange *[fig. 50]*.

— In describing SymbioticA and our residency program, I would urge you to first check our website: www.symbiotica.uwa.edu.au.

— As for the residency program – maybe the best way to introduce it is to take some passages from an introductory document we prepared for potential residents (not currently on the website). If you read between the lines you will most probably realize that we needed to address some issues that were raised in our seven years of hosting artists in biological labs and our attempts to maintain the type of research culture we developed:

— *SymbioticA is a unique organisation that has developed many programs, one of which is to provide great opportunities for local, national and international artists to work in a scientific setting. We have regular queries from potential residents e very week and this is set to increase given the growing profile of SymbioticA.*

— *Ideally we would like potential residents to visit SymbioticA for a short period to see what facilities are available. The first short visit is intended for artists to refine their project after assessment of the availability of local resources and discussion with scientists and artists in SymbioticA. If this is not possible, please research the web to familiarise yourself with the University.*

— *SymbioticA realizes that a Science Faculty may be a foreign environment for most artists. The School of Anatomy and Human Biology has numerous possibilities and potentials for collaboration. We recommend you look at www.anhb.uwa.edu.au and the University of Western Australia site www.uwa.edu.au to familiarize yourself with the experts you may wish to collaborate with.*

- *A search for special research interests can be done at: www.directory.uwa.edu.au/specialists/*
- *A residency at SymbioticA should be for a period of at least three months and can be up to one year in length. Projects can take on a variety of forms; it is important to remember that SymbioticA is a research laboratory and not a production studio. This, however, does not exclude you from developing a finished work as part of your residency. Whilst Directors may approve of the initial project proposal to be researched regarding its feasibility - there is the expectation that the time in SymbioticA may result in the project diverting from its original course.*
- *SymbioticA is a small organization and is able to provide working space, access to resources and other laboratories within UWA, introductions to UWA staff and WA organizations relevant to your project, but much of what you do as resident must be self-motivated. While we support your project with as much 'in kind' support as we can, SymbioticA cannot subsidize your project in any financial capacity. We request you ensure that you have adequate funds to cover your project and your own living expenses during the residency. For the ongoing funding of laboratory activities, expensive laboratory consumables must have been budgeted for. Please consult with SymbioticA regarding the preparation of your budget for funding bodies.*
- *Every Friday at 3pm SymbioticA has open discussionson various relevant topics around the science, arts and humanities. The forums are open to anyone interested. As a SymbioticA Resident you would be expected to attend as your perspective and participation would be beneficial to other residents and visitors.*
- *There are safety rules that must also be followed within a science department. Information on this is found within the student guide at www.safety.uwa.edu.au.*
- *Whilst SymbioticA is able to give our support towards applications, this does not automatically mean the University of Western Australia Ethics Committee will grant approval to your project. SymbioticA Directors will guide you in applying for ethics clearance for your project. This must be sought prior to commencing your residency and information about this can be found at:*
- *Human ethics:*
- *http://www.research.uwa.edu.au/ethicsacu/welcome/ethics_animal_care/human_ethics*

- *Animal ethics:*
- *http://www.research.uwa.edu.au/ethicsacu/welcome/ethics_animal_care*
- *SymbioticA aims to foster collaboration and to facilitate unhindered exchange and developments of ideas. We advocate the ethos of Open Source and Creative Commons.*
- *As part of the University of Western Australia, SymbioticA follows the University IP guidelines as found at http://www.legalservices.uwa.edu.au/welcome/copyright/features.*
- *SymbioticA has adopted the Creative Commons ethos of the accumulated knowledge generated in SymbioticA being shared and credited amongst researchers and residents. The copyright on the artwork is retained by the artist; however, there are still some uncertainties in the legal and artistic communities in regard to whether the outcomes of some artistic biological processes can be copyrighted. Projects researched and developed during a residency should acknowledge scientific and artistic collaborators and SymbioticA.*
- *As a new resident you will have access to the cumulative knowledge already gathered by SymbioticA researchers. You are expected to contribute to this common knowledge pool by providing an acquittal, which includes documentation (of protocols, procedures, lab book), an artistic report on your project, financial summary, and any comments about difficulties and suggestions for SymbioticA's future residents. SymbioticA perceives this as an important aspect of your residency as we are regularly approached by academics, journalists, and other artists regarding our programme of activities, and this acquittal ensures you are represented in our archive. It also ensures other artists coming into the space become aware of the process other residents go through. We request that the acquittal is received within three months of completion of the project.*
- *Applications are assessed quarterly by the Directors of SymbioticA. Specialists and appropriate SymbioticA artists in residence are called on should their expertise be required in the assessment. Should you become a Resident, you may be called on to assess, in confidence, other applications. There are no deadlines for application, but please note that it may take up to three months to process your application. This is in order to ascertain each applicant's project and feasibility in relation to all the other applicants for this period. In many cases the application will be returned for revision based on the comments of the directors and external advisors.*
- *Send your application including all of the above to e-mail: sym@symbiotica.uwa.edu.au*
- *Or mail to the address below:*

— *SymbioticA Residency*
SymbioticA
School of Anatomy and Human Biology
Mailbag M309
The University of Western Australia
35 Stirling Highway
Crawley. Western Australia, 6009.
The projects are assessed quarterly and are selected based on:

— *Quality of concept*
— *Quality of past work*
— *Innovative nature of the project*
— *SymbioticA's ability to contribute to the project.*
— *Suitable funding being sourced by applicant.*

JILL SCOTT

— Immersion in the lab is revered in art and science as one of the most valuable ways to transfer knowledge from educational facilitator to student. According to [Silvia] Caravita, the lab context in any discipline is so fundamental for learning and for the exchange of information that education is problematic without it.[135] Immersion for artists in the science laboratory is an excellent starting point for new educational approaches to transdisciplinary practice. Caravita also suggests that a scientist must actively build knowledge through the personal interpretation of her/his experience, but must share this experience not only with peers but also with "outsiders."

— Why do you think that the artist can become this very valuable "outsider"?

— Generally, from our experience with the life sciences, the most beneficial way to start working with scientists is to assist them to collect the empirical evidence they need for their research. For example, in the artists-in-labs project at the Centre for Microscopy, artists were given a "hands-on" education on tools like the Scanning Electron Microscope. The mastering of skills like these requires a very exact level of mimicry as part of the learning process.

— How important is it for the artist to really understand the methodologies and methods of science, and why?

ORON CATTS

— The hands-on approach can be read on different levels, and artists are approaching it in different ways. From my perspective, getting into the lab does not always entail getting one's hands wet.

— I once made up a taxonomy of models of artistic engagement with the life sciences and came up with twelve non-exclusive ways (I will not attach names to the categories, but I am sure that you will know the types...):

1. The illustrator
2. The commentator/representer

3. The visitor/guest/onlooker
4. The appropriator
5. The entertainer
6. The user
7. The industry worker
8. The hoaxer
9. The hobbyist/amateur
10. The after-hours/under-the-table
11. The mail-order/ready-made
12. The researcher/embedded in science/technology setting.

— I am particularly interested in the last model, but deliberately inserted the somewhat loaded term "embedded" in it, as I think, that in the same way that the integrity of reporters embedded in military (such as in the war on Iraq) should be questioned, so is the case with artists. Saying that, my own experience, and observing more than thirty residents coming through SymbioticA and "getting their hands wet," it is important for some nonbiologists to enter the life science lab and engage with the manipulation (not just visualization) of life in the most direct and experiential way (literally, the phenomenological way...). The knowledge one gains by the multisensorial experience of dealing with life in such a way is something that neither text book nor image can provide. The major thing that SymbioticA provides our residents with is getting equitant with the tools and techniques of the life science (this is not the same as science making...) so they can research their projects (and in some cases produce the work) themselves.

— The notion of the outsider is valuable to some degree, but it is not my main interest. Is the embedded reporter in the war on Iraq also a valuable outsider? Is it about the artist providing fresh perspectives to assist the scientist in her pursuit? My answer is that this is problematic.

— Some examples infer hierarchy in the artist/scientist relationship. For example, artists collecting empirical evidence needed by scientists for their research may not be the best (or most beneficial) way to start working with scientists. For me, it seems that it cements the position of the artist/researcher as inferior to the scientist. In SymbioticA, we appoint our residents as honorary research fellows in our school (Anatomy and Human Biology). This position puts them on equal footing with the other researchers (scientists) they are working with. The relationships that are then formed are those of mutual mentorship at the start and colleagues as the project continues. It is not so much about being in the service of the other, although in some cases we get some of the scientific collaborators complaining about the fact that they feel used. This is more to do with the different ways the art and science worlds treat egos.

— I think that in the context of artists working with living systems, it is not just important for the artist to understand the tools of science, it is imperative for the artist to gain the skills that will allow her to do the work herself. However, following scientific methodologies is something I recommend our residents not to do. In my experience, it is really important to delineate to some degree the ways our artistic researchers operate from the way the scientists work. It is not about artists doing science or scientists doing art. My experience shows that in most cases when this is being attempted, the result is bad science and not so interesting art.

RICHARD WINGATE

— There is indeed great deal of value in a constant explanation and defense of ideas and results with the nonscientist, the "outsider." Best when it's with a creative mind that has a similar interest in observing, interpreting, and understanding the world (artists, playwrights, children, sometimes students). However, I'm not sure that the artist in collaboration is necessarily an outsider. Over the last year or so of hearing Andrew Carnie talk about my work, I feel he invests it with more interest and excitement than I can muster. He explains the concepts equally well and is an excellent advocate for science and scientists. I think, incidentally, that this is something that he has conveyed in some of his postings. So Andrew's an "insider" but not a bench research scientist.

— Could he be? Well, the training is long and hard—there are degrees and doctorates to be won along the way—and he's already been through his own long and hard professional development, with so many parallels in structure and benchmarks for achievement. Andrew, what do you think?

— So when a nonscientist is trained to use a scanning electron microscope (Jill's observation), the use of the term "mimicry" is very accurate. Superficially daunting equipment can be used efficiently and correctly by anyone with a relatively brief training. But these are just tools. The instinctive understanding of the significance in form and function—what to discard and what to leave in—is gained through experience and often a great deal of scholarship.

ANDREW CARNIE

— I do wonder whether our training sets our brains differently. I wonder whether the situation is analogous to sport. In sport, preparation sets up the body for the rigors of the particular activity. It is impossible to think a thin-framed, lean cyclist like Lance Armstrong could be involved in the Tour de France, and in between be an American League football player in the Super Bowl.

— The world I think my work comes from as an artist is the area of slippage and leakage between brain areas, the fortuitous accident, and the ability to recognize its value. Living at ease with this state, and maintaining it, is difficult. In science a different rigor is needed. It is strange how many artists are dyslexics, and I wonder if this is part of the brain set needed to be artistic. I did start training to be a scientist and think I could have done this; but I think much of either of these disciplines is dammed hard work and doing the amount of labor twice over to be fully trained in two disciplines is hard to conceive of unless the economics of matters changed massively to help free up lots of time. My dyslexia would certainly have got in the way, and I could never do the pared down writing that Richard Wingate achieves for his journal submissions; that I am jealous of.

— Richard, the artist's studio is often only occupied by one person, but this does not stop it being the site of many a love-hate relationship. The hate when one thinks, why did I ever start on this painting, why did I ever begin on such a precarious career; to some, great self-satisfaction and self-love on the occasion of the completion of some work which has ended up far beyond the expectations to which it was conceived, and for that moment when one is alone in the studio to love, the space, and the chance to begin to play with materials and ideas.

— Back to your other posting on "hands on" and "outsiders"; I very much agree with the sentiments of what you say, it seems the conversations that have crossed between us, as we have been observing, interpreting, and understanding the world, have been the most important moments, and this just emphasizes my wish that all disciplines become more permeable to "outsiders."

SUZANNE ANKER

— To Oron's nameless "taxonomy of models" of artistic engagement with the life sciences, I would like to add a few others:

13. The moralist
14. The activist
15. The careerist
16. The "no talent"-ist
17. The justifier
18. The didactic-ist
19. The salvationist
20. The alarmist
21. The narcissist
22. The "wannabe"-ist
23. The whore de culturist.

— With twenty-three pairs of human chromosomes in tow, I think I will stop here. To this Duchampean model of types, [136] Oron brings into public view the very serious issues embedded in live laboratory practices. As phenomenological and sensorial experience, working with life forms in real time gives one pause about the complexity and fragility of living systems. And as Oron introduces elsewhere in his philosophical texts, he brings attention to such concepts as the "semi-living"[137] and "aesthetics of care."

— Tissue engineering, bioprinting, and bioscaffolding remain on the edge of the ways in which the body can be conceived of as harboring its own repair kit. At a time when the formerly abject (umbilical cords, circumcision tips, etc.) garner renewed value, the future has arrived.

— Additionally, I am interested in knowing more about laboratory practices as experienced by scientists and/or artist/researchers. Vladimir, can you tell us more about the research you are engaged in? Where does the idea of bioprinting come from?

VLADIMIR MIRINOV

— Our goal is to print living human organs suitable for clinical implantation. Our most ambitious project is the Charleston Bioengineered Kidney Project.[138]

— Organ printing is basically the biomedical application of well-established, rapid prototyping technology. Suzanne Anker was probably one of the first artists who successfully employed rapid-prototyping technology in her art. The difference in our approach is that we do not print plastic scaffolds or biodegradable temporal supporting frameworks and then seed them with living cells in a bioreactor. It is a so-called two-step approach. We print living cells and biodegradble stimuli-sensitive hydrogels simultaneously. It is a so-called one-step approach. Bioprinting of living tissue of desirable geometry is definitely a new form of sculpture, or even better, say, biosculpture. The principal difference is that this sculpture is living sculpture, and it can grow, evolve, and undergo change and remodeling. We can use as building blocks living cell aggregates or clusters of different colors. In such an approach it can look like a three-dimensional painting. The painter Paul Klee wrote that painting is putting the right color in the right place. Similarly, bioprinting can be described as putting the right cells into the right 3-D space. In this context "biosculpturing," especially when one will use cell aggregates of different colors, can be called "three-dimensional pointillism."

— The idea of using rapid-prototyping technology or computer-aided, layer by layer additive manufacturing originated from our frustrations and unsucessfull attempts to put living cells into porous 3-D biodegradable scaffolds. These scaffolds required a high level of cell density as well as enormous precision. We do not grow organs, as many wrongly believe. We are assembling living tissues and organs from cell aggregates using a bioprinter or robotic computer-aided cell dispensing device. Living cells have a unique capacity to self-assemble into 3-D living human tissues. We just put them in the right place at the right cell density. Cell viability and tissue fusion permit the supporting biodegradable hydrogel, thus allowing printed cell aggregates to form. We then use accelerated tissue maturation technologies in order to finish the job and transform the printed viable tissue constructs into living 3-D tissues and organs. It is interesting that a cell can undergo cell sorting processes or preferential adhesions. Thus an artist can use flat petri dishes or flat perfused minibioreactors with projectors, and place cells with differing colors (green or red fluorescent proteins) to observe the dynamic living paintings, or biopainting. By projecting the image on the screen or wall, cell patterns evolve and change with time. Evolving cell and tissue biopatterns can be as beautiful as flowers. There are no technological problems in creating a minibioreactor with projector. Biopainting can be a very popular art if the artist or painter has bio-ink of differing colors. Already there are identified and analyzed types of genes that are responsible for the expression of florescent proteins. These genes, after being introduced into the cell, can express seven types of fluorescent colors. All these tools are waiting to be explored by creative artists.

— Furthermore, one can ask if one day we will be able to realize Pygmalion's dream and bioprint as a biosculpture a complete living human being. Only the future can answer this provocative question. I am afraid that human printing technology (if one can imagine the logical possible social implications) can induce the hottest political debate in the future. Recent debates on stem cells are only the beginning. I think artists can help to promote these inescapable and forthcoming debates at the present time.

— The bottom line is thus: living bio-art, both in two-dimensional biopainting (living dynamic biopatterning) and three-dimensional biosculpturing (bioprinted 3-D dynamically evolving living tissue of desirable shape) is already technically possible.

ANDREW CARNIE

— Within my artistic practice, I have had the experience of visiting many labs. However, I have not spent much time working in the lab alongside scientists. When I did work with Richard Wingate, on one particular occasion I do remember rather messing up the "staining" I was carrying out of chick embryos. I figure it was just lack of practice. On most visits to see Richard, it had been to spend lengthy sessions in discussion on the topics Richard has mentioned in his previous postings. There has always been talk based around the theme of science and much talk on technical matters, video editing, and computers. During the course of these talks, visits would be made from his office to the long-shared lab of the developmental neurology unit, part of which he occupies with his research team. Progress on particular projects was viewed from prepared slides or by looking at new material through the confocal microscope.

— In part, I also fear spending too much time in any one particular lab, being engaged and absorbed too much. I wonder how many interesting and strong works I could necessarily make embedded in a lab for a long period, say anything over a year. I have often been looking for the particular essence of the science I am approaching, the idea, and sometimes this is quite quickly gained. Further time in the lab might be wasted in my case. However, in general I would love to see more integration between the sciences and the arts, as I would, in fact, for all fields of study. What I have been able to do is to swap and move between collaborators, and this jumping about has its benefits. It has brought refreshment and new ideas to the artistic endeavour each time. En route, I have worked with neurologists and geneticists. Currently I am working with a neuropsychologist on temporal lobe epilepsy and in another project with a heart transplant team and with Margrit Shildrick, a bioethicist.[139]

— One big defining aspect in this research process is the actual time spent on any resulting artistic work. Making the work I produce is often very labor intensive. The piece I completed for the Art and Mind Festival, Winchester,[140] last year included the production of 648 images. This time-based slide dissolve piece, *We Are Where We Are*, occupied six months of fairly solid labor *[figs. 51, 52, 53, 54]*. This is not an uncommon amount of time to spend on any one work that involves photography, drawing, and endless computer manipulation. It is obviously an aspect controlling the amount of time I could spend in any lab. The reverse is true for scientists. Finding time to discuss their work is always very difficult. Richard Wingate has always been very generous, and we continue to meet to talk despite the completion of our projects a few years back.

— In my visits to science labs there are elements of practice that I have seen that I would like to bring back to the studio. Some of the organization has been impressive. My own applications to funding bodies have always been sporadic, but I liked the organized manner in which I saw scientific teams expect they would always be making applications. I certainly like the experience of the labs I have been in, often quite open-plan in nature. There seems to be a good deal of interchange between professionals. Artists tend to be quite lone individuals, isolated from others. The common sense of purpose in Richard's lab has always seemed comforting. More minds working towards one goal is exciting. Since experiencing these various labs, I have been much keener on joint projects and working within interdisciplinary teams. The heart project with Margrit Shildrick mentioned above involves three other artists, a sociologist, and philosopher.

RICHARD WINGATE

— I don't think of the laboratory as a creative space—it's more like a battleground. It's a place where we wrestle with temperamental equipment and discomfort and the site of almost constant defeats and only the occasional heart-stopping victory and revelation. It's also a collegial atmosphere of constant banter, gossip, humor, and even flirtation. Because so many people pass through and invest each inch of bench surface with the story of their struggle, the laboratory is often redolent with memories and evidence of past colleagues. It is a very "human" (and I'm conscious of the conversations in this session) environment, not at all impersonal. Scientific creativity is, in my experience, exercised on whiteboards and paper, in offices and seminar rooms, on the commute into work, and at home in the late night around the kitchen table.

— The studio, by comparison, seems enormously liberating—a playground. However, I'd imagine that what the studio shares with the laboratory is a love-hate relationship with their respective occupants. Is this true?

SUSAN SQUIER

— I hate to leave this virtual conference in the middle of such ferment: love the idea of the virtual plush pets (though of course I worry about the Gundification of resistant science studies), and wanted to spend some time talking with Miriam van Rijsingen about the chicken art of Koen van Mechelen. I did know his work, have been following it for a couple of years, though the website does seem to be a dead-end for the last year or so. (Has he switched websites?) There are a couple of artists working on chicken breeding, which, in the worst-case scenario, seems to me to fall in the category of "bespoke animals," and to produce a queasy sense of undifferentiation between the poultry industry and the art industry.

— But I'm off to a zone with no e-mail, so I'll wish all of you well and look forward to checking in at the end of the symposium (if it's still up on the 18th) to see what has emerged. I do hope some of those graphic fiction writers have checked in, perhaps to talk a bit about the pluripotentiality of graphic fiction images, in contrast to the specificity of biomedical imaging.

GIOVANNI FRAZZETTO

— As a practicing scientist, I feel summoned to first address the question introduced by Suzanne Anker concerning the problematics of being a scientist. In one sentence, I think the main problem for scientists is the difficulty in coherently mixing three main activities, that of knowing, doing, and being. In other words, how much of their knowledge and practice of their profession clashes or in general coexists with their way of being in the world. Is a scientist's field of expertise influencing his/her world view and actions?

— In his 1918 lecture "Science as Vocation," Max Weber states, "Natural science gives us an answer to the question of what we must do if we wish to master life technically. It leaves quite aside or assumes for its purposes, whether we should and do wish to master life technically and whether it ultimately makes sense to do so."[141] Let us go one step further: is science, more strictly biology, also a means of self-clarification? Is biology the way to true being? Michel Foucault stressed the inseparable and problematic link between subjectivity and truth in Western culture. How intrusive is the subject in this process of objectivication? Does this truth define me or tell me who I am?

— There is no doubt about the pervasive power of new discoveries in biology on society and on individuals. Biological knowledge is employed to define individuals in biological terms. For scientists, who are at the same time producers and consumers of this knowledge, the identification with it is more problematic. The question is whether science practice is an ethical activity that can be ascribed to a pattern of conduct, a *lebensfuehrung*, or self-regulation. For instance, the dissection of the neurochemical mechanisms underlying emotions brings forward a more "disenchanted" vision, of which scientists are the producers.

— Having read the collection of scientists' journals edited by Jon Turney and having kept a journal myself documenting the evolution of my daily experimentation, I can hardly agree more with the conclusion that scientists are as human as anyone else. This type of self-writing (as opposed to experimental notes or laboratory records) promotes "reflexivity" and traces the intersection between knowing, doing, and being that I introduced at the beginning.

— Coming back to the issues presented by Carl Djerassi at the start of this session—I agree that narrative and theatrical representations can certainly be very powerful means to describe scientists' dilemmas or idiosyncrasies. All the plays mentioned, and many others, do this exquisitely well. However, I agree with Suzanne Anker when she reminds us that the public perception of scientists is not necessarily equal to the vision scientists cherish of themselves. Also, I am not sure I can agree with the statement that hidden among lines, graphs, and pictures of a laboratory notebook are often burning personal issues. When generating results, an experimenter's personal issues do not cease to burn or exist in parallel. However, they are supposed to remain invisible to the reader of those lab protocols. Detailed historical analysis of objectivity (Lorraine Daston,[142] Peter Galison[143]) speak to the rise of an "a perspectival objectivity" as the ethos of the interchangeable observer and operator, unmarked by style and idiosyncracy that might interfere with the generation and communication of results.

SUZANNE ANKER

— Marvin Heiferman has curated several exhibitions centered around "artworks that are science based." Particularly known is *Paradise Now*, a seminal exhibition on this subject which he co-curated with the late Carole Kismaric. In addition to its venue at Exit Art[144] in New York, the exhibition traveled to many other university art galleries and museums within the United States. Pamela Auchincloss facilitated these arrangements, and the exhibition's outreach was impressive.

— It is interesting to note that this show received an enormous amount of press coverage: the *New York Times, The New Yorker, Science News*, etc. There were also TV interviews and other e-media postings. What was curious to me at the time (and still is) is that the critical response from the art world per se and its allied agencies of cultural consensus were, if not totally absent, then at least frail in their capacities to engage this subject. The major art magazines, *Art in America, Artforum, Flash Art*, etc., did not write any significant feature articles on this conceptually engaging, internationally driven theme. What accounts for this erasure is speculative at best. Certainly many art historians and cultural theorists, from Barbara Maria Stafford, Donna Haraway, Christiana Paul, Mark Dery, W.J.T. Mitchell, and James Elkins, to cite a few, are significantly ensconced in this dialogue.

— The questions at hand address a possible incompatability between current institutional frameworks and experimental art practices. The notion of art's epistemic value is presently comatose within the art world's established hierarchy. At a time in which art consumption is on overload, collecting art has become the latest trend in cultural cachet. As the writer Anthony Hayden-Guest has so aptly stated, the art fair has become "the new disco."[145]

— Marvin and Nancy, what are your thoughts on these matters?

MARVIN HEIFERMAN

— Getting back to the issue of audience and the role of art-sci in the art world, I'm happy to read in posts that artists are finding their own venues and making their own opportunities to show works and to generate dialogue. As Suzanne suggested, the experiences around the exhibition *Paradise Now: Picturing the Genetic Revolution*, which opened at Exit Art in New York in 1999 and traveled for a couple of years, and critical responses to it helped me to understand that the art world is not necessarily the best place to go to have a discussion about images, their meaning, and impact.

— Yes, the exhibition got a fair amount of attention; it was a novel project we were lucky enough to have some foundation money to fund, market, and promote. What was interesting to me about *Paradise Now*—and particularly in its first and full iteration in New York—was the opportunity to see how art and science could coexist in an exhibition space—whether and how, for example, scientific data and information could be passed along to viewers; how, for instance, displays of genetically modified foods could be juxtaposed with artworks speculating about how similar products might impact our lives, the environment, etc. The exhibition included objects including the packaging for a porn tape entitled *Designer Genes*, one of Dr. Watson's original models for DNA from the early 1950s, images from science fiction films, bumper stickers from organic farming groups, as well as very provocative artworks that directly or not so directly alluded to the relationship of genetic research to race, eugenics, and profiteering. The more than forty artists in the exhibition ran the gamut from scientists-turned-artists to conceptual provocateurs to artists who aestheticized science. And that, I think, is what made the exhibition interesting. Exhibitions that have tackled similar subject matter since seem, to me, to have taken a more predictable, one-sided, accusatory approach to the subject matter, genetics is bad/art is good.

— I had hoped the exhibition would trigger discussion. And it did. And it didn't. It did among the art-sci crowd because the project was big, visible, eclectic, and gave credit to the artists who had been working, unheralded, in the area for years. The exhibition proved enormously popular for school groups—lots of students came, from elementary- school-age kids to groups of interns from NYU Medical School. The exhibition space was almost always crowded. One of the most interesting events was an evening in which artists from the exhibition met with research fellows at Rockefeller University. After a few drinks some lively discussions kicked in, and some interdisciplinary collaborations that started that evening are still ongoing.

— The show got lots of press, but what it didn't get was "serious" attention in the art press. To the art world, the exhibition seemed like a novelty event, perhaps a little too in-touch with the real world for comfort. "Serious" critics had a hard time getting serious about the work in the show. Because much of the work dealt directly (sometimes analytically, sometimes fancifully) with pressing, specific, and sometimes disturbing issues, I think it was easy for some critics to dismiss the art as being illustrative; which is not to say that some of it wasn't. But with over three dozen artists, there were a lot of history, ideas, and images (some pretty spectacular) flying around. I suspect that one problem I saw in the art world's response to this, and to similar projects, is that the art world likes to set and control its own agenda, and to select the issues that it thinks are important. And for the most part, those issues—while they are often provocative, engaging, pleasurable, critical, whatever—are not often, or terribly, consequential.

— The art world gets its power by creating a little distance between itself and the real world. I think the science part of the art-sci equation makes people nervous. Which is why (in addition to logistical issues) the exhibition, when it traveled, was pretty much stripped of its scientific and popular culture components. It became, more clearly, an art exhibition. And even so, it tended to travel not to stand-alone arts institutions, but to university galleries, where it engendered interesting interdisciplinary programming. But I still can't forget what it was like to sit on a panel at Carnegie Mellon, where the show traveled, and listen to the director of a local contemporary art museum who just dismissed the whole art-sci endeavor—the engagement of dozens of artists in scientific images and issues—as silly, and never felt the need to explain why.

— For me, the opportunity to work with scientists on exhibitions has been, and continues to be, eye-opening and rewarding. I'm working on a major exhibition project now for the Smithsonian about how photography has not only been used by, but has changed every discipline that makes use of it. I've been lucky enough to meet and talk with astronomers, natural historians, physicists, historians from the Air and Space Museum, keepers of the earliest known daguerreotypes of the moon. The images and the conversations are incredible. I feel lucky to meet and work with people in the sciences, where imagination and thinking routinely create and bounce off of images. This brings me back to the art world. The commercial art world is, despite its own hype, pretty parochial. And really, most collectors and/or museums don't want to hang what they believe are scary pictures up over the couch or the admissions desk.

— The point is that images of all sorts that are being made in laboratories and in the course of research are (a) interesting and revelatory to look at and (b) may actually change the way we understand our lives (all of our lives) and the world we live in. I don't think those are the kinds of ambitions that fuel the art world of the twenty-first century. That said, I know there's lots of artists out there with a genuine interest in the issues and images of science, and I support their interest and their work. I also applaud the dedication and entrepreneurial spirit it takes to create venues to show that work and opportunities to discuss it, like this conference.

[public blog]

STEVE MILLER

— Technology has given us another career, maintaining e-mail and websites and surfing the blogsphere. With that in mind, please forgive any repetitions, should this posting be another mirror of something already said. It's been difficult to find the time to follow the entire discussion. This subject is vast and time has limits, so I have been lurking and putting the demands of an exhibition ahead of blogging. In defense of my online slacker attitude, I chime in with a few personal observations

— One of the things that always drew me to the subject of art and technology was that this visual vocabulary is a new language system that represents our time. The opening statement of this conference makes that clear as well. Artists used the science of perspective in the Renaissance and freed themselves from Byzantium. I'm not sure the language of science will become a commercial aspect of mass culture or release us from it or anything else. However, it is the computer (a scientific development) and its special effects that decorate mass culture and help create our spectacles. The film 300 just earned 70 million in its opening weekend. The technology of Google, My Space, You Tube, and Instant Messaging is the new pop culture. Those examples may not be science, but the Internet is the new international media, and the Internet is building a massive audience and its favored billionaires. On a smaller scale, the fact is that this Internet blog allows for an impressive global participation. Nonetheless, science still sits outside the mainstream.

— Marvin Heiferman asked, for me, the most provocative question: who is the audience for this online debate and for the images or words created by scientists, artists, playwrights, historians, and social anthropologists in this world of sci-art? Of course, WE are: those who are participating in the conference evidence its value. No doubt, there are probably as many answers to the audience question as the number of participants. It should be noted that I have often had conversations with several contributors to this conference about audience. Who is the audience, and how much energy is spent trying to further develop this audience?

— It was interesting to see from Marvin's observations that, with the *Paradise Now* exhibition, a significant part of the audience was non-art world. Paradise Now provoked a discussion. It is curious that this discussion took part in academia more than in the art world. As Marvin mentioned, for some art critics, lots of the artwork came off as pseudo-science, a one-liner, looking like the illustration of an idea, or aestheticized science. As a participant in that exhibition, I may fall into some of these categories. Yet in spite of my aesthetic roots, for this particular project, I made a "scientific" investigation. I extracted a collector's blood, replicated their DNA, photographed their chromosomes under an electron microscope, and silk-screened these images onto painted canvas. It was my effort to redefine the traditional art category of portraiture. Access to new technologies in 1993 allowed me this privilege. Working with science permitted an investigation inside the body on a molecular level. The result was not science, but perhaps worked better as metaphor. Science now allows us to look inside the individual, and the work we need to do is an inside job.

— Scientific research and artists working in the lab with scientists is still an important part of my process, so that was an interesting topic of discussion for this conference. Part of my practice has included working with Dr. Stephen Adler at Brookhaven National Labs, where Dr. Adler was writing code for the Relativistic Heavy Ion Collider, as well as with Dr. Rod MacKinnon, the 2003 Nobel Laureate in chemistry. His research involved understanding how a positively charged ion moves across a cell membrane. The point of all of this is to recognize that even though I go to the lab for my sources, my goal is to make art. It's ridiculous to think that I am a scientist or know enough about science to make a contribution to that body of knowledge that is specialized and specific. However, there might be some artists who come from a position in science so that they might actually make a contribution to that base of knowledge.

— Art is about something else, another form of knowledge. Some art today pushes hybrid forms where, amazingly enough, a glowing green bunny can be genetically conceived and the material of laboratory-grown skin can be a new canvas. It's incredible and stimulating to think that these materials can be a new medium for art. Maybe it's not so surprising. If Warhol can make mass media a new medium, the field of art is wide open for any variation, ranging from the revelations of science to the cultural obsession with fashion.

— Back to that nagging question of audience. It's an important question. Matisse said, "Audience completes the work" (or something like that), and part of this conference is about art. The technical aspect of science makes a particular demand on the viewer. As someone else noted on this blog, some people found this artist's images of brains repulsive and scary. In 1993, many people said my portraits reminded them of death or a negative experience they had in the hospital. As Marvin said, some of the issues surrounding at and science just make people nervous.

— A recent tour of five New York art fairs certainly proved that: this year, the art-sci movement has not made the commercial art fair circuit. Evidently, it takes the P.T. Barnum skills of Matthew Barney (the mutant) or Damien Hirst (the pharmacy) to get aspects of the art-sci debate to a larger audience. For many years Ashley Bickerton has been raising the issue of surviving the environmental apocalypse. These three examples are noted because we live in a complex world where the concept and its dissemination are not mutually exclusive, and one of the ways to create audience is to get the art to market. It's not the only way—this panel is another. One can argue that a larger audience share can dilute the message, but some artists, like Warhol, seem to have it both ways. Can the art-sci crowd?

— For the most part, artists working in this area of debate will show their work in exhibitions. Any art exhibition that is driven by a strong agenda includes plenty of art that gets selected to fulfill the mandate of the agenda, but is not necessarily good art. While the art-sci agenda is vast, the category may be too narrow to reach a larger audience. This may be especially true in the current art world that appears to get much of its energy from the power of market forces. Art-sci is successful as an impassioned niche (as seen in this conference) and the popularity of *Paradise Now*. The proof that this audience is expanding is that the online conference has been extended past its original termination date! Perhaps the size of the audience is a red herring rather; it's the opportunity for content-driven work to communicate.

— In the end, art is a very different animal than scientific practice, and the goals are different. Some artists will always be interested in science, and that can create a crossover genre and an interesting hybrid. More importantly, science does offer an attempt to escape the "conspiracy of art" recognized by Jean Beaudrillard in one of his last books. His premise is that the homogenization of aesthetic language makes the visual styling of Target and Starbucks not much different than the many good-looking, well-made cultural products available throughout the world. Within this universal "sameness," the symbolic value of art is lost. Foucault looked in the margins to define the center and understand civilization. For precisely because of its nonmainstream status and with its specialized language, science provides a meaningful visual medium for art. It may be that the issues of science are fascinating to discuss, but it's the artist who must rise to the occasion.

MARCH 11–13

SESSION

3

SOCIAL AND CULTURAL IMPLICATIONS OF VISUALIZING THE BIOSCIENCES

SUZANNE ANKER

— In 2003, the fiftieth anniversary of the discovery of the structure of DNA, James Watson, accepting an award at the New York Academy of Sciences, talked a bit about the residual after-effects of his seminal scientific findings.[146] He explained that after the paper he co-authored with Francis Crick in *Nature* in 1953 not much of anything happened.[147] Although DNA's structure had been exposed, its application to the fields of the biological sciences was still uncharted territory. Fast-forward fifty years plus, and alternative dilemmas are upon us.

— The plasticity of the DNA molecule itself makes it possible to slice and dice, recombine, insert, or knock out gene sequences at will. One can conceive of the biological and technological sciences as forming a free-floating, yet diaphanous backdrop, intervening into the activities of human life. From the food and the medicines we ingest to the tests that screen our interior bodies to its use in criminal investigations and social hierarchies, DNA technologies continue to alter the ways in which we inhabit the world. This compelling technology, like others before it, is Janus-faced and relies heavily on our ability to assess risk and its data, models and their interpretation as well as an understanding of the underlying principles in bioethics *[fig. 55]*. In this session, our discussion will focus on the social significance of the genetic revolution.

— Artists and writers, and especially sociologists, philosophers, and ethicists, continue to play catch up from a humanist point of view in assessing the uses and abuses of these fast-paced technologies. How shall we manage the social consequences of scientific facts?

— Each discipline encompasses its own set of practices with regard to this question. From your particular perspective, please identify the issues, values, and ethics involved in the current range of transformative bio-practices.

TROY DUSTER

— In a later posting, I will respond more specifically to Suzanne's question about core issues, but this post is to address an underlying assumption in her query: "How shall we manage the social consequences of scientific facts?" And indeed, are the rest of us simply "playing catch-up"?

— Today's (Sunday) *New York Times* has two pieces on the opinion page that are germane to this discussion. They both refer to the recently announced discovery that the different national groups of the United Kingdom and Ireland are much more closely related genetically than previously thought. One of the authors uses this as the occasion to suggest that such research has the potential of uniting us in our common humanity—bursting the bubble of ethnic and racial essentialism that is often the source of warfare, strife, and of course, a long-standing rationale for oppressive social domination.[148] The other author documents the history of that domination of the Irish by the English, using literature and quotes from essayists of the eighteenth and nineteenth centuries to reveal just how deeply embedded was the idea that the Irish were an inferior race. [149]

— The problem is that social domination is not dependent upon demonstrated genetic or biological similarities or differences — but upon the capacity of one group to control the resources, politically, culturally, economically, and of course, militarily. My case in point is a study, now nearly a decade old, that compared the DNA of Arabs, Jews, and Welsh subjects, and concluded that the Arabs and Jews shared much more of the DNA-patterned markers with each other than either did with the Welsh. To say that these findings did not much "budge" or destabilize patterns of control or domination in the Middle East would be a wry rhetorical understatement. In a similar vein, the study of the DNA of the priestly line among Jews was used to try, at the laboratory level, to establish a group in sub-Sahara Africa as one of the lost tribe, and thus "essentially" Jewish. Many things come to mind about the impact (or lack of it) from the study, but "the primacy of biological ties above all else" is definitively not one.

— The purpose of my bringing this to the fore is to counter the notion that the biosciences are somehow in the lead, or at least should be in the lead, in determining what really constitutes human genetic variation and diversity, and if the rest of us just followed that lead, we would come to some better understanding of the human variation among us—which until enlightened by genetic analysis, is flawed, imprecise, and certainly not scientific. We may well be informed that Strom Thurmond, Thomas Jefferson, and countless others dipped into another gene pool (read this as socially framed, not genetic), but the Thurmond and Jefferson families have no more "budged" than have the Israelis, the Arabs, the British, and the Irish—and the reason is simple enough: genetic analysis will not adjudicate policies that sustain long-standing practices of human social and cultural stratification.

— More in a bit on core issues.

JILL SCOTT

— I don't necessarily want to put a spanner in the works, but one of the things I would like to bring up in relation to this section is the obvious gender imbalance in the hard sciences and the problems women artists might encounter in such environments.

— In Europe, as elsewhere, it seems that entirely new creative approaches as well as relational and societal impact assessments are sorely needed to entice women to become more interested in the hard sciences. Last year, our artists-in-labs project team conducted a gender survey with Arts Catalyst[150] in London and the Art and Genomics Centre in Amsterdam.[151] Differing attitudes towards creativity seem to be the cause of an enormous gender imbalance in many Northern European countries, especially in biotechnology, computer science, physics, and engineering departments.[152]

— Of course, in relation to science and gender discourse, science writers in the US are leaders in the field. For example, in Evelyn Fox Keller's book *The Century of the Gene* (Harvard, 2002), she proclaimed that the few women who are engaged in genetic research always provide a much "more creative social approach."[153] The science question in feminism has also been raised very thoroughly by Sandra Harding. Harding clearly states: "Perhaps we should turn to our novelists and our artists for a better grasp of what we need, because they are professionally less conditioned than we (scientists) to respond point by point to a culture's defenses of ways of being in the world."[154] She claims that artists who acquire solid information about science bring very sensitive issues to the public for scrutiny.

— Currently we are in the process of forming ARTSACTIVE, a worldwide art and science network,[155] which can explore solutions to the issue of shared creativity by training more artists in science. Perhaps we can also harness the potentials of transdisciplinary practice to involve women in more creative approaches to science. If mature women artists were trained in scientific fields, could they produce mediated art and design works that emphasize the creative potentials of scientific inquiry? Could these artworks then be distributed to art departments in secondary schools, where alternative role models are sorely lacking?

[public blog]

ROGER MALINA

— Jill,

— Let me pick up on your gender question, which is part of a much deeper one about how scientific laboratories and science institutions in general are "connected" to the societies that they are part of.

— Staying on the artists-in-labs topic, the UK arts council, with Leonardo co-sponsorship, has placed five artists in the Space Sciences Lab at UC Berkeley.

— Of the five artists, four were women (one was a wife-husband team).

— The lab itself is dominantly male, but not the science education group that served as "broker" for the residencies.

- The selection juries were gender balanced.
- It is my observation that in the art and technology community in general there are numerous outstanding women artists and scholars—yet when they work with the science instutions, we are faced with organizations that are dominantly male, especially in the management levels (and tenured faculty).
- It would be naive to think that these cultural diffrerences don't affect the way that art-science collaborations work in laboratory contexts.
- And in the biosciences so much of the discussion bears on the nature of life and our responsibilities towards life forms that as we talk about visual culture and the biosciences, there must be gender differences in approaches.

SUZANNE ANKER

— Discourses on gender and society abound in the cultural and scientific literature. In this symposium there have been several references to social issues regarding gender identity and its power politics. As new technologies reconfigure the sexual revolution into an asexual one, Susan Squier's initial comments about trans-sexuality and reproductive rights target an aspect of this circumspect and unchronicled territory. Lee Silver in *Remaking Eden* goes into full detail about the conceivability of male pregnancy.[156] He also cites a time when human embryos may be created from the fusion of cells from same-sex parents.

— Brad Davis further engages the ways in which the female is erased by the imaging practices of sonograms. These images focus only on the fetus itself, as if it is located somewhere else. The new technologies of sex selection raise ethical dilemmas concerning the pervasive practice of aborting female fetuses, particularly in China and India. Recently the American College of Obstetricians and Gynocologists (ACOG) released a statement on the nonmedical use of sex selection as a sexist practice.[157] However, this Committee Report challenges debate with regard to the nature of an individual's reproductive rights. What I find so compelling in this symposium is that as soon as it began, reproductive rights and embryonic images flooded in. In what way are we engaged in updated variations concerning gendered anatomies? Or on the other hand, is the fetus a "primal marvel" that still remains opaque to us? Are there any sociologists out there? Richard Twine and Troy Duster, what do you think are the core issues regarding gender, society, and technology at this point in time?

CATHERINE WALDBY

— Here is an excerpt from a forthcoming article of mine and Melinda Cooper's about female reproductive biology as an investment site for regenerative medicine and biocapital:[158]

— *Throughout the OECD [Organization for Economic Cooperation and Development], birthrates are in decline. Women in the majority of the developing nations are delaying childbirth and having fewer children, a trend that has precipitated considerable anxiety among states concerned with the dwindling proportion of their working populations and the complex economic and political consequences of an aging citizenry. The reasons for this shift in reproductive priorities are complex, intertwined with transformations in the biopolitical ordering of life.*

— *These transformations could be summarized as the neoliberalization of life, both in the sense of the everyday life of citizens and in the biological life of populations. The decline in reproduction demonstrates these two forms very succinctly.*

In the realm of everyday life, we see the effects of a shift in state-market-citizen relations. The postwar form of state-centered biopolitics[159]*– national health systems, social security, Keynesian full employment policy, and economic regulation —gives way to what is sometimes termed the Competition State*[160]*—concerned with the attraction of finance capital, the deregulation and privatization of production, and the devaluation of its work force to achieve global competitiveness. The Fordist model of family life (male breadwinner, family wage, full-time mothering) has necessarily given way, in the face of deregulated wages and the need for the two-wage family, to the financialization of everyday life,*[161] *the decline in social security, and the increasing cost of health care and housing as they are opened to global investment markets. These changes dramatically increase the economic and emotional costs of reproduction, and lead women, especially middle-class women, to delay childbearing or avoid it altogether. The wide dissemination of feminist-influenced civil society also means that state exhortations to have more children are unlikely to find much purchase.*

— *It is evident, then, that one of the unintended consequences of neoliberalism has been the state's loss of traction over female reproductive biology and its disengagement from nation-building projects.*

— *At the same time, women's reproductive biology has become the focus of extensive biomedical research interest and global commercial innovation. This constitutes another form of neoliberalized life, this time situated at the level of biological processes, and part of a much larger marketization of biological vitality.*[162] *Effectively, we would argue, the processes of reproduction have been deregulated, privatized, and made available for investment and speculative development. This investment takes two major forms. First, since the birth of the first IVF baby in 1978, medically assisted reproduction has become a huge global business.*[163] *Middle-class couples increasingly turn to Assisted*

Reproductive Technology (ART) (IVF, donor gametes, PGD)[164] *to facilitate conception late in a woman's reproductive life, once they have achieved economic security. Increasingly, access to ART and donor gametes is through reproductive tourism, the purchase of fertility from poor women in the developing world.*

— *Second, and more recently, many of the new technologies associated with regenerative medicine—embryonic stem cell research, savior siblings, somatic cell nuclear transfer (SCNT),*[165] *cord blood banking—rely on female reproductive biology as a generative site. These technologies utilize the autopoietic capacities of embryogenesis and the fetal-maternal blood system to generate therapeutic stem cell tissue that is itself autopoietic.*[166] *That is, unlike whole-organ transplant, which substitutes a working organ for a faulty one, regenerative medicine aims to transplant tissue that is self-organizing and self-generating, once inside the body, able to repair and regenerate diseased sites. These technologies effectively convert the generative power of female reproductive biology into regenerative therapy. Hence, they position reproductive biology as one of the most important machines for the bioeconomy—especially as a promissory machine, working through appeals to biological potential and the future regeneration of the body.*[167]

— *Female reproductive biology is thus undergoing a complex rearticulation. New reproductive technologies like IVF have disaggregated it from its in vivo location, and stem cell technologies have diverted it into biomedical domains unconcerned with the production of children. Reproductive potential is now bifurcated. In vitro embryos and in vitro oöcytes can be transplanted to produce another human life, a child; and they can be biotechnically reconfigured in a laboratory, diverting their pluripotency into the production of embryonic stem cell lines.*[168] *In both cases, however, reproductive industries require proprietary control of high volumes of difficult-to-donate reproductive tissue, either to supplement the failed fertility of the IVF patient or to perform the tricky task of creating stem cell lines. Hence, the compliance, negotiability, and general agency of female populations is a central issue in the development of the reproductive bioeconomy.*

RICHARD TWINE

— I would say that the main issues remain some of the classic concerns from feminist science studies: the masculinity of scientific practice—several participants have responded positively to the assertion that scientists are as human as the rest of the human race (sorry if I've misquoted that, but it's come up a few times). Presumably, then, there is a context here in which this is a reaction against assertions of the *in*humanity of scientific practice. Of course, the issue suffers partly from a meaningless generalization, but remains in tension with long-standing arguments about the privileging of various values: the disavowal of empathy, mastery of nature (and the body essentialized as nature), anthropocentric and utilitarian ethics, and so on. Obviously then, as I think Jill pointed out, the issue of gender is of special interest in art-sci interactions. Are there new exclusionary relations at play in terms of which artists are involved in these projects and so on? To what extent is gender an object of exploration in art-sci projects?

— The other main issue, that has already been well expressed in Catherine's post, concerns new modes of harnessing biovalue across reproduction, and the consequences this has for women (and indeed the nonhuman female) and social/familial relations generally. Participants have already spoken about the sociotechnological construction of fetal citizenship, and we can point to other changes, such as the partial technological challenge to heterosexism. Generally, we ought to tie in our contemporary analysis to historical work[169] and think through continuities/discontinuities within the classic modernist project vis-a-via the gendered biopolitics of bringing reproduction under "rational" planning and control.

EUGENE THACKER

— What's interesting about this is that the oöcytes become this reservoir of "biovalue" (to use Cathy's term) — a kind of surplus that can be transformed "vertically" into a viable human life, as well as "horizontally" into a range of nonhuman biological entities (e.g., stem cell differentiation).

— This seems to open onto other sets of questions as well. Without taking us too far from Jill's questions, I wonder if this neoliberalization also relates to a new type of labor —one different from the maternal body and the attendant technologies involved in reproductive biology. What "labor" is performed by cells, eggs, and embryos? It sounds a bit ridiculous, I know (no, I don't mean embryos as a form of child labor in Chaplin's factory). But much of the science behind these innovations is predicated on the notion that, given the right conditions (e.g., the right growth factors, etc.), a group of cells will themselves differentiate in a particular way—with a minimum of intervention. Of course, there is a lot of intervention. But the concept is that the biological entities will do it themselves. There's an interesting approach of "pulling back" here that seems consonant with the neoliberal "flexibility" surrounding reproductive technologies.

— Another, related question relates to Richard's comments about biopolitics and the modernist project of the rationalization of life. There's a new kind of biological "life" here that is neither natural nor artificial—even the language seems less about top-down instrumentality but more about bottom-up "control," "perturbations," "coaxing." If this is so, is this also the case in earlier examples of rethinking the concept of "population" in terms of the management of biological flows, fluctuations, and averages?

CATHERINE WALDBY

— Thinking about the productivity of biological material as a form of labor is really useful. Effectively, this is the source of productivity of the bioeconomic industries—their ability to tweak in vitro and in vivo living materials and turn their normal productivity into labor for biocapital, as it were. At the same time, I would want to link this labor to what I would call human in vivo labor. Many more areas of biotech involve ongoing relationships to populations who give access to their in vivo biology as well as their ex vivo tissues—biobanking, where banks take blood samples but also monitor the lifetime health of their cohorts. Clinical trial participants, increasingly drawn from third world populations where this is their main source of income, women sell ova on a regular basis to the reproductive industries and now to stem cell research. The last two of these are very onerous and involve significant risk and often pain. In each of these cases the labor performed by tissues requires a prior and ongoing labor performed by subjects. I think the biotech industries would like to escape this relationship and rely purely on in vitro life, but in many arenas they can't.

VLADIMIR MIRONOV

— Population aging is not biological but rather a social phenomenon. Unjustifiable transgenerational transfer of economic wealth through [Otto Von] Bismarck's introduction of a nonpersonalized pension system is the main reason. Bismarck was afraid of socialism, and life expectancy had a longevity of fifty years during this time.[170]

— Many religions are against abortion because abortions reduce the number of churchgoers. In former Japanese villages, old generations were forcefully eliminated, which effectively prevented population aging. According to Swedish studies, retired people live on their earned pension money only the first five years of retirement, and then the younger generation supports them (transgenerational economic wealth transfer). Economically enslaved youngsters have only two choices: either be poor and have children, or enjoy life and do not have children. The older generation must live on their earned money (personal pension account) and/or help youngsters to raise children. Old Japanese methods are not socially acceptable, but economically driven de-population can be considered as some form of human designed populational genocide. Pension systems are considered a great social achievement of economically developed countries. But as usual, "The road to hell is paved with good intentions." The

American Association of Retired Persons (AARP) is one of the most powerful lobbys in the USA Congress.

— "Young people of all countries, unite!" (I am fifty-two years old; maybe at sixty-five I will reserve my rights and somehow change my opinion.)

CARL DJERASSI

— I am bothered by the direction taken by some of the messages and comments dealing with "reproduction." Some seem to deal quickly and superficially with a vast number of issues and do so rather devoid from the real world. The following sentence is such an example: "As new technologies reconfigure the sexual revolution into an asexual one, transsexuality and reproductive rights target an aspect of this circumspect and unchronicled territory."

— What sexual revolution is being talked about? That of the 1960s, dealing with sexual behavior and its impact on the power relation between men and women, or the reproductive revolution starting in the 1980s, created by IVF technologies, with a very different impact? The above sentence only makes sense if sex and reproduction are always bundled together. As I have written in books, scientific articles, novels, and plays, what we are facing is a separation between sex and fertilization. Sex, as always, for love, lust, fun, curiosity, or whatever, and fertilization is increasingly under the microscope. This is happening primarily in the "geriatric" countries of Europe, Japan, and soon also the USA, where the average family consists of 1.5 children. Planned fertilization and consequent reproduction now happen so frequently that they must be firmly separated in our discussions about sexual behavior.

— There is plenty of realistic and societally crucial new territory, foremost of which is the postponement or circumvention of the biological clock, because that impinges on the lives of millions of women and has an enormous effect on the power relationship between the two genders. Sex predetermination falls into this category and so do many of the implications of preimplantation embryonic genetic analysis. But to put potential "male pregnancy" into the same pot is absurd. It is a form of mental masturbation that, as most masturbation, is enjoyable, sterile, and basically harmless, except when it offers fodder to the lunatic fringe and reproductive fundamentalists. I am, of course, not referring to [Lee] Silver as the lunatic fringe, but rather to some people who will make horror scenarios out of this. Face the fact that enabling male pregnancy is of no priority whatsoever among reproductive scientists and is of so little importance, globally speaking, that it isn't worth arguing about, as compared to the enormous implications of the other applications of reproductive technologies.

— The Ars Electronica 2000 Festival of Art, Technology, and Society on NEXT SEX in Linz[171] offers a good example of what I am worried about. During that excellent and wide-ranging festival, Nobuya Unno[172] from Japan reported on his work on the artificial placenta, which included stunning video footage of four-month-old fetal goats inside an artificial placenta, where they were hooked up to a dozen or more tubes, pumps, and artificial feeding systems and thus kept alive for several weeks. Since they were already at an advanced stage of development, they moved around to a startling extent and seemed "live." Of course these fetuses were removed from the goat uterus and placed into the artificial placenta when they were already fairly advanced in order to see whether they could be brought to maturity.

— The ostensible ultimate aim of this research is to see whether even twenty-two-week-old, super-premature humans could be kept alive and allowed to mature for another few weeks, so as to provide viable babies. Without now arguing whether such efforts are really worthwhile from a societal standpoint, at least the scientific rationale was made clear. But the journalists present at that conference all focused on the question of whether this meant that in the future women would be able to have babies in that fashion by depositing three or four-day-old embryos into such artificial placentas and then pick up the intact baby nine months later. Aside from the preposterously difficult scientific hurdles to be overcome (tubes and pumps can be hooked up to an advanced fetus but hardly to an embryo or blastocyst) and the unbelievable cost of such an artificial placenta baby, is it really worthwhile to seriously debate this issue? Given the existing opposition to societally useful applications, is it worthwhile to offer fodder for horror scenarios to the strident opposition? It does far more harm than good, and the same applies to discussions of male pregnancy. A nine-month ex-utero gestation or successful male pregnancy is not the societally important issue of this century.

SUZANNE ANKER

— Let us for a moment turn our attention to other sentient creatures from current bio-ethical and historically mythical perspectives. From the Paleolithic inscriptions of bison on cave walls to the "biofacts"[173] in our laboratories, our relationship to the nonhuman animal continues as an open question. Recently, there have been many texts and exhibitions on this subject, including our relationship to animals in zoos and circuses. Nato Thompson's exhibition *Becoming Animal: Contemporary Art in the Animal Kingdom*, at MASS MoCA [Massachusetts Museum of Contemporary Art] in the US, included the work of some twelve artists exploring the animal/human connection.[174]

— Conceptual artist Mark Dion's work approaches the subject of the animal by critiquing institutional and environmental contexts in which they are inhabitants. Drawing from his research in natural history, taxonomy, and archeology he posits his art practice in ecological and philosophical terms. For filmmaker Kathy High, two rescued laboratory rats, which she names Echo and Flowers, become protagonists in her film *Embracing Animal* (2005). Injected with human DNA in their embryonic state, Echo and Flowers were transgenically fabricated as research tools. Biologically driven, yet strategically man-made, these fabricated beings can be considered under the nomenclature "biofact."[175] Catherine Chalmers's photographs of animals point to culturally determined hierarchies imposed by humans on the animal kingdom. In her work, she has employed the lowly cockroach, which she herself carefully bred to act out ironic scenarios *[fig.56]*. Although the big cats such as lions and tigers are, in effect, genetically programmed killing machines, they have a much more revered status in society than lowly rodents or insects.

— In what ways do critical fictions, a.k.a the cultural imaginary, function as rhetorical sources for the creative imagination in both studio practice and laboratory immersion *[fig. 57]*?

LEONEL MOURA

— Mistreatment of nonhuman life forms is based on anthropocentrism and its discourse of human superiority. And we don't need to go very far to learn the reasoning of that kind of "philosophy." For example, Vladimir Mironov, a panelist in this debate, on an earlier post stated that we humans are the "most sophisticated species on this planet." The conclusion that he delivers in his paper is as follows:

— "What makes us humans? Humans, according to definition, are not animals. One can call humans bio-social creatures, but what makes us really different from animals? The answer is very simple: humans have something that animals do not. Animals do not have religion, art, science, and technology. Shortly speaking, animals do not have culture."[176]

— Humans are not animals? Curious. Biosocial? What about ants? Although it is not clear if (nonhuman) animals have religion, which to be true would be very fortunate for them, or science, which would be a temporary setback in evolution, it is evident that certain animal behaviors and skills can easily be considered as art and technology. Anyway, the fact that some people, and a lot of them, need to demonstrate that we are unique and superior just reveals a profound and dangerous antagonism in regard to life in general. This is the same antagonism that generates abuse, exploitation, and slaughter of all nonhuman life forms. Edward O. Wilson, Richard

Dawkins, and Stephen Jay Gould, just to mention a well-known few, explain brilliantly this process.

[public blog]

URI

— The idea that we are special compared to other animals is an absurd one. We are not special, except possibly in the multiplicity of scale that we can interact with the world and with our capabilities of abstraction. My proof of this (for myself) has always been that many other animals can play games, i.e., pursue an activity that does not feed them or aid in their survival and which requires imaginantion. Cats and dogs do this; birds do this (grey parrots for one); many types of ape and monkey do this, etc. As in the example of ants, any specialty of human ability has other animal examples. However, it is not because we are special that we use other animals. It is simply in our nature: (a) we are toolmakers; manipulators both of the conservation of animals and the husbandry of animals are the kinds of things we do; (b) we are omnivores and so eat meat. I believe it is within our best interests to husband the planet as best we can and not kill things or waste things. On the other hand, human culture can not afford to stop studying the world to rest on its laurels, and therefore we must continue research; otherwise we face stagnation and, eventually, cultural disintegration. This invaribly means animal research as we have not as yet reached the edge of biological understanding to do such resarch without studying actual animals.

VLADIMIR MIRONOV

— From a biological point of view, anthropocentrism is just an "evolutionary competitive advantage of the human as a biological species." From a social perspective, belief in the equality of human and nonhuman life is basically a return to the oldest primitive forms of religion. I mean, an unchallenged belief in Mother Earth can also create absurd scenarios, such as dying of hunger rather than killing animals or engaging public lawyers to defend ants and insects.

— In Soviet Russia, during the Nazi occupation of Leningrad, people ate not only all possible animals, like dogs and cats, but even other human beings in order to survive. So animal killing and eating is not so much a question of philosophy as it is sometimes a question of species survival. I think from a position of anthropocentrism (which is the essence of all modern dominant forms of religion), people must first eliminate human hunger on the whole planet and then, and only then, start to talk about animal rights with a reasonable sense of morality and social responsibility.

— Saving animal lives while ignoring the demands of human life is probably a most counterproductive idea, both from a scientific, evolutionary point of view and from the position of most popular religions based on anthropocentrism. I assume that more advanced religions provide an evolutionary advantage to humans as species. Thus, going back to a more primitive form of religion is to regress and is not progress. Finally, many forms of animal rights activism in the extreme are an illegal, uncivilized, and unacceptable form of social behavior. I think it is much more moral to help people in Africa who are dying from hunger on a low protein diet by providing them with affordable animal meat, than to teach them how to be vegetarians.

— I could not imagine a person who will call an animal rights specialist when they see rats in their house. In this situation people usually call for deratization experts. So what is all this fuss about blocking legal mandatory experiments on rats and other animals during drug testing and drug toxicity? Illegal activities against mandatory federal rules and regulations established by a democratically elected government are not democratic. It only increases costs and delays the drug discovery processs. As a result, people are dying waiting for new drugs. Additionally, I do not see substantial and sufficient investments from the multiple, well-funded animal rights organizations into the development of drug toxicity assays. These assays are based on using human cells or in silico models of drug research.

— Bioprinting technology can help to fabricate such mini-organs and 3-D human tissues based on pharmaceutical developments. Instead, university scientists are forced to spend time and money on unnecessary, consuming paperwork for the protection of the research lab from animal rights criminals. If animal rights organizations want to be consistent and not hypocritical, they must invest in animal-free tissue engineering meat technologies[177] or human-cell-based drug experiments. I do not see this happening, at least on a sufficient and necessary scale. Empty hypocritical rhetoric and illegal criminal activities are not the way civilized people must deal with these issues. Artists who promote animal rights views associated with criminal activities undermine evolutionary anthropocentrism and promote a return to more primitive religions. We must all be socially responsible for our actions, both artists and scientists.

LEONEL MOURA

— I don't wish to polemicize. It should be interesting to see if there are distinctive perceptions on this issue among artists and scientists.

— E.O. Wilson wrote, "Congo, who deserves to be called the Picasso of the great apes, was responsible for nearly 400 [paintings]. Although most of [Congo's] efforts consisted of scribbling, the patterns were far from random. Lines and smudges were spread over a blank page outward from a centrally located figure. When a drawing was started on one side of a blank page, the chimpanzee usually shifted to the opposite side to offset it. With time the calligraphy became bolder, starting with simple lines and progressing to more complicated, multiple scribbles. Congo's patterns progressed along approximately the same developmental path as those of very young human children."[178]

— I see many similarities with my artbots, in particular with the first ones *[figs. 58, 59, 60]*.[179]

SUZANNE ANKER

— Many posts in our conference have referred to questions about "the animal." From a companion species (Haraway[180]) to living test tubes (Karafyllis[181]) to food sources, our co-evolution with nonhuman species continues to raise innumerable questions, both ethically and psychologically. At a time when H.G. Wells fantasies of Dr. Moreau move out of the laboratory and into social space, questions concerning "taxongenomic crash," as I term it, rise to the forefront. Examples include the accelerating ease with which diseases are jumping from animals to humans and Darwinian selection between industrialized animals and "wild-type" species. Martin Kemp's forthcoming text, which is cited here, historically traces human thinking and picturing in regard to our sentient others. I invite you to a preview.

— Martin Kemp, *The Human Animal in Western Art and Science* [182]

— *If it is dangerous to make man see too much how he is like the beasts, without showing him enough of his grandeur, just as it is to make him see his grandeur without showing him enough of his beast-like qualities, so it is even more dangerous to let him ignore one or the other. (Blaise Pascale, 1632-62)*

— *Part I: Humours, Temperaments and Signs*

— *Chapter 1: Fixing the Signs*

— *Chapter 2: Feelings and Faces*

— *Part II: Souls and Machines*

— *Chapter 3: From Meaning to Mechanism*

— *Chapter 4: Fable and Fact: La Fontaine and Buffon*

— *Part III: Going Ape*

— *Chapter 5: Beastly Boys and Admirable Animals*

— *Chapter 6: Our Animal Cousins*

— *Chapter 7: Art and Atavism*

— *Introduction: Facing up to ourselves*

We all exhibit a propensity to react to members of the animal kingdom as if they have personalities that we can read from their appearance. We also manifest a tendency to see individual people as bearing some kind of resemblance to one of our fellow vertebrates, or even invertebrates. The title of this series of studies is to be understood in this double sense – humanised animals and animalised humans. There is rich historical legacy of imagery, which illustrates both sides of this animalistic coin.

— *On one hand we have the wide vein of animal representation, ranging from compilations of natural history to the illustration of animal fables. It would be easy to disassociate the more obviously humorous and story-telling aspects of the animal capers from the "serious" scientific illustrations. But this is certainly not valid historically. In the first great printed picture-books of animals in the Renaissance, the character and meaning of the animals as defined in the "Bestiaries" was of as much concern as what we might regard as more scientific data. And fables, for their part, could serve serious philosophical ends. It might seem the whimsy of legend was progressively expelled from the portals of zoological science, above all in the 19th century. However, the element of "story" in the natural history of animals has continued to exercise a powerful if somewhat covert hold on the imaginations of those who aspire to build up a picture of nature in action. Darwin is a key participant in this respect.*

— *At the more popular level, we can all recall as children the central role that animals played in our developing consciousness of character, behaviour and life-narratives. Children's story-books are replete with speaking animals, often standing on two legs. How many of us have not possessed beloved cuddly or less cuddly stuffed animals, accorded human names and very distinct personalities? It is striking how relatively constant are the virtuous and the villainous amongst the child's cast of animal characters. Teddy bears – improbably, given, the fierce irascibility of bears in the wild – have become the most popular repositories of warm feelings and trust companionship. Over the longer span, we expect to find dogs and cats and small birds featuring strongly in the lists of the basically good. Snakes figure hardly at all amongst the virtuous. Foxes, amongst the canines, are not regularly regarded with feelings of trust, whereas badgers, which are at least as keenly carnivorous, generally fall on the friendly side of the divide. Wolves are generally beyond the pale. Certainly for Red Riding Hood. Though for the Romans a she-wolf suckled their legendary founders, Romulus and Remus, and there were persistent legends of wild children nurtured by wolves.*

— *Clearly the actual behaviour of the animals in relation to humans plays a powerful role in our instinctive perceptions of their characters. The most sustained relationship that we have enjoyed with any animal is with the horse, the noble beast that for generations carried men into war, pulled wheeled vehicles and tilled our fields. Only exceptionally does the horse play anything other than a worthy role. Equally, it is difficult to imagine a dairy cow or woolly sheep acting as the villain of the piece (or peace), even less so if they feature in their youthful guises as calf or lamb. Bulls and rams are a different matter. Gender and age are obviously potent factors. On the other hand, some notably violent and unsociable animals regularly play admired roles. Our cherished cats are noted for their ritual slaughter of mice and baby birds. Even the lion is conventionally granted a noble personality, according to a legend of many centuries' standing. There is hardly a building erected by European and many oriental rulers that do not parade lions as representative of the potentates' just but fierce virtues. Tigers occupy in a more ambiguous position, desirably signifying the high performance of Esso petrol, while not commonly regarded as promoters of peace and harmony. Bats, creatures of the dark, are not much liked, in contrast to many birds. Black crows do carry sinister connotations, however, while black and white magpies are notorious thieves.*

— *Inhabitants of the waters are best considered on a case-by-case basis, but tend on balance to fall on the distrusted side of the emotional spectrum, with sharks and octopuses regarded as particularly malevolent. Fishes' eyes tend to be regarded as unappealing and unsettling, and we often refer to something "fishy going on" when we suspect that deceit is being perpetrated. Amphibians and reptiles also tend to evoke feelings of revulsion. To describe someone as "reptilian" is certainly not a compliment, and we may suspect that they are likely to be involved in fishy or creepy activities. The aesthetics of touch undoubtedly play a role in stigmatising such genera of animals. Wetness, sliminess, hardness and temperature all have roles to play. Someone lacking warmth of emotion is described as a cold fish.*

— *Non-vertebrates as a whole tend not to get a good billing, though snails clearly do much better than slugs. Animals with angular exoskeletons exude an air of aggression, reminiscent of knights in armour. Strangely, such "medieval" insect garb has become one of the stock in trade features for futuristic sci-fi warriors from alien planets. Insects and crustaceans are not fondly regarded as a group. Butterflies are an exception, even when masquerading as caterpillars, as are bees for the most part, but not wasps (unless you own a Vespa scooter in Italy). Spiders can sometimes be regarded as OK, especially with respect to their arachnid weaving skills, and little "Money Spiders" can be children's favourites, but they more often stimulate strikingly adverse reactions beyond rational explanation. Even in Britain, where there are no native poisonous spiders, the sight of a big, black specimen who has ascended the waste pipe into a white bath is likely to produce a shudder of dread. Size and hairiness are definitely significant factors in defining a spider as threatening.*

— *Ancient legends of the animals, particularly as recorded in "Bestiaries" told of idiosyncratic habits that were frequently read in terms of human meaning. The legacy of the legends is found in popular imagery and sayings, "Hiding one's head in the sand", as ostriches were wont to do, has come to symbolise not facing up to reality. When we manifest insincere grief, we are "crying crocodile tears". The key compilations of fables, by Aesop and La Fontaine, including many illustrated editions, and the recurrent tales of Reynard the Fox, have typically dressed up salutary lessons for humans in the garb of picturesque stories of talking animals. Reynard's legendary cunning is both a literary topos and related to the well-documented inventiveness of foxes pursued by hunters. As in other territories inhabited by the human animal, many beasts retain relatively constant characteristics, such as the kingly lion, though even he could be satirised as occasion demanded.*

— *Over the ages and across cultures, many different animals have been regarded as sacred, and even as deities. Just to take two examples, monkeys are sacred in Sri Lanka, while cows are accorded divine status in India. The term "sacred cow" has entered common usage to indicate something or someone who cannot be criticised (often with the implication that such status is not altogether warranted). The many Egyptian statues of cats testify to their divine attributes, while combinations of feline and aquiline characteristics in such compound beasts as griffins signify origins and powers beyond those of natural creatures. Generally speaking, creatures fantastically assembled from the component parts of diverse species have served as harbingers of divinity or devilry. Almost all religions that have exploited figurative imagery have developed fantastically confected agents for good or for evil. Human-animal compounds carry a particular frisson, as mermaids, sirens, harpies, centaurs and satyrs testify.*

— *The perceived character of particular animals, male and female, real and imaginary, is not subject to simple generalisation, and once we embark on the assembly of lists it is difficult to know when to stop a complex and often unstable mesh of cultural factors, knowledge, experience, and deep-seated instincts are involved, collectively and individually. The same creature can play the roles of polar opposites. Many people love birds of all kinds. A few cannot stand to look or touch of feathers. And one of the scariest of all Alfred Hitchcock's masterpieces of filmed unease was "The Birds", in which flocks of flapping insurgents wreaked mass revenge on humans, as fronted by the lovely Tippi Hendren. The complexity and fluidity of our feelings does nothing, however, to diminish the elemental power of our reactions to each creature in particular circumstances.*

— *The power of a film such as Hitchcock's, addressed to adult audiences, shows that whatever we may have sloughed off amongst our childish things, we retain a very strong instinct to anthropomorphise animals and to build them into our own stories. Domesticated companions are obviously in the front line, but no visitor to a zoo is likely to the resist the pull to assign character on the basis of appearance. Even the popular names of some wild animals and, most notably, birds speak of their fancifully assigned roles in human society. The Secretary Bird, for instance, is all neat precision and high-heeled pertness.*

— *Orwell's exemplary "fairy tale" of revolutionary animals taking over their farm, and the inexorable onset of hierarchies and divisiveness, gains much of the efficacy of its characterisation from our instinctual anthropomorphising. The initial assembly was convened by "old Major, the prize Middle-White boar", now venerable and stout, but still "majestic looking". Of the sketches of the animals, all of whom behave as more-or-less true to type, none are more finely characterised than the horses, the beast best known to humans over the ages:*

— *"The two cart-horses, Boxer and Clover, came in to together, walking very slowly and setting down their vast hairy hoofs with great care lest there should be some small animal concealed in the straw. Clover was a stout motherly mare approaching middle life, who had never quite got her figure back after her fourth foal. Boxer was an enormous beast, nearly eighteen hands high, and as strong as two ordinary horses put together. A white stripe down the centre of his nose gave him a somewhat stupid appearance, and in fact he was not of first-rate intelligence, but he was universally respected for his steadiness of character and tremendous powers of work… At the last moment, Mollie, the foolish, pretty white mare, who drew Mr. Jones's trap, came mincing daintily in, chewing a lump of sugar. She took her place near the front and began flitting her white mane, hoping to draw attention to the red ribbons it was plaited with." Writing this, as I am, in Los Angeles on the night that the Oscars are awarded, I sense that Mollie would had felt more at home amongst the bejewelled starlets than with the pigs snuffling in their troughs. Though in Orwell's tale it is the pigs that are the cleverest of animals.*

— *At the very end of the book, the pigs are to be discovered forging a self-serving alliance with neighbouring humans. Looking into the house at the summit meeting of pigs and farmers, "Clover's old dim eyes flitted from one face to another. Some of them had five chins. Some had four, some had three. But what was it that seemed to be melting and changing?…" Later, following an argument between Mr. Pilkington and Napoleon, the leading*

pig, who had been accused of cheating at cards, "Twelve voices were shouting in anger, and they were all alike. No question, now, what had happened to the faces of the pigs. The creatures outside looked from pig to man, and from man to pig again; but already it was impossible to say which was which."
This last line of Orwell's novel flips to the other side of our coin; that is to say, to our parade of animalised people.

VLADIMIR MIRONOV

— Returning to the core question concerning the alteration and manipulation of life forms, I hold that evolution is based on the alteration of biological forms and DNA. It is a fact.
If we want to stop evolution, we must try to block it; if not, we must rationally deal with this. Agriculture was always based on genetic manipulation, breeding and selection. Transgenic plants and animals are nothing new from this point of view, just a continuation of an ongoing technological and evolutionary process. The main reasons why people are afraid of new technologies can be attributed to misleading media hype, inferior education, lobbying activities, and unscientific propaganda of special interest groups. We must not be afraid or block these changes but rather learn how to socially accomodate the new emerging technologies.

— The genetic revolution is dramatically, but also strongly, overhyped. The FDA approved an overhyped gene therapy that still does not exist, at least in the USA today. There are no FDA-approved drugs based on overhyped genomics discoveries. The sad truth is that after the adoption by BIG PHARMA, all these overhyped genomics, proteomics, metabolics, system biology, and other –omics, FDA approval of drugs was reduced in number. It will probably take several decades until society will get all the benefits from the genetic revolution. It takes time to build a village. The potentially most dangerous consequence of the genetic revolution is designing of a novel type of biological weapon using synthetic biology. In case of catastrophe or war, novel bioweapons could eliminate the human population, or dramatically reduce it. But humans can, must, and will survive.

— I think scientists could not be a part of lobbying groups for "micromanaging" the social acceptance of specific technologies, because there is an obvious conflict of interest. The job of scientists is to provide maximally objective information about potential social outcomes of any evolving technology. The whole society, not just an educated elite with special interests, must decide on what is socially acceptable and what is not for a given specific society and historic time. Religious leaders in a theocratic society also have doubtful legitimacy. The record shows a very unimpressive endorsement of dealing with rational scientific and technological innovations. Religious authorities tried to block the development of anatomy as a science, which is the basis of modern surgery. I did not read of any apology about this from any religious leaders, who meantime have no ethical problem in using modern advances in anatomy-based surgery. Thus, I think that the whole society, rather than special interest lobbying groups, must decide how to proceed.

— Mandatory public disclosures of conflicts of interest, public debates, and democratic approaches should aim for transparency and accountability. Dealing with the social consequences of scientific facts must be maximally evidence based, not belief based. The truth is that science is highly undemocratic, elite-based, and a meritocratic enterprise. However, if science could not provide objective evidence and predictive outcomes, then common sense or decisions based on common values are the best ways to manage the social consequences of scientific facts. That is why religion or ideologies which provide common, shared values for specific societies will never die. The history of technology teaches us that Luddites are usually losers.

— If something is technologically possible, and if there is enough social demand for this technology, then usually nothing and nobody can stop this technology from eventual implementation. The key words are: "social demand."

— From my perspective, these are the social issues involved in biopractices: embryonic human stem cells versus adult stem cells; modifications of germ cells; possible changes in human identity after organ or stem cell replacement; human "chimerization" using animal cells or xenotransplantation (transplantation of living animal tissues and organs from transgenic animals to human); tissue-engineered brains; cloning and/or bioprinting of complete humans on demand

— Finally, I would like to propose a futuristic scenario of a possible outcome of the genetic revolution:

— Human embryos and fetuses which are acephalic, that is without heads, are exempt from the legal regulations of late-term abortion.

— In somatic nuclear transplantation, an adult cell can be transplanted into a denucleated human embryonic stem cell.

— Gene therapy can induce acephalic genes into the embryonic stem cell.

— These genetically modified embryonic human cells can be implanted into a surrogate mother, who has an already in-place legal contract. These cells also would be autologus and would allow for no immune rejection.

— If society will accept and permit the genetic modification of germ and human embryonic stem cells, we could technologically create acephalic embryos.

— Persons without brains technically would not be a person so there is no ethical problem.

— If acephalic bodies can be called a human being, than organs like kidneys or groups of organs would also have to be termed a person. Of course, this is obvious nonsense.

— Question: Can we use acephalic technology to grow autologous embryonic human organs for clinical transplantation? Will such technology be socially acceptable? A leading UK embryologist even suggested the use of animals as possible "surrogate mothers" for such acephalic human embryos.[183] This is not my idea, I am just communicating it.

— Bottom line:

— We will never escape from new ethical, social, and political challenges created by emerging technologies. I think the role of artists is to formulate these problems and controversies in dramatic ways so as to make the general public aware of these concerns. The job for scientists is to provide maximally objective evidence and logical sequences for implementing the emergent biotechnologies and biopractices. In order to make rational judgments and informed decisions, the general public and society as a whole must have the right to objective information from differing sources.

SUZANNE ANKER

— In an earlier post, Catherine Walby referred to the term "autopoiesis." Leonel Moura cited the term "genetic algorithm" in relation to his image-making robots. How can these "systems" concepts be integrated into our discussion? What vernacular language can be employed to demystify these terms? How do they relate to signification and embodied knowledge?

[public blog]

VANGELIC

— In response to Suzanne Anker's post on autopoeisis: As an architect, I am wondering how Francisco Varela's and Humberto Maturana's research on autopoeisis, cognitive science, and neurophenomenology could intervene in current art discourse?

— Currently, some architects are beginning to pierce the static Vitruvian model of space. If we expand the notion of space within a systematic framework, space takes on a very different meaning. Francisco Varela's book *Embodied Mind* (MIT Press, 1991) describes some astonishing experiments. The experiment Varela describes in his book clarifies his concept of what he called "enactive knowledge." After many years of research, he discovered that there was a certain type of information gained through "perception-action interaction in the environment." In many aspects the "enactive knowledge" is more natural than other forms, both in terms of the learning process and in the way it is applied in the world. "Such knowledge is inherently multimodal because it requires the coordination of the various senses," he wrote.

— Analogously, in addressing the complacent pluralism in the production craze towards representation, concern with "what" we represent instead of "how" we represent, are we relying too much on our visual perception? Varela infers that this is the point of departure in how we relate to others and the world-perceptions combined with movement in time. And this can only occur when we can have a sense of a distance to be moved through, not of a distance separating us from time, where the social merges with the existential. Modern physics has forged a widespread understanding of the interdependency of time and space. Space is not a given but is created in time by the movement of bodies and sound waves, many scientists have affirmed. Space is an unstable, time-based concept that has to be established by our perceptual and cognitive apparatus through continuous looking, listening, and moving. Thus we construct our spatial environment through perception (especially seeing and listening) and proprioception (the intuitive spatial "knowledge" of our body; for instance, in eye-hand coordination). This also means the relation between space, movement, and body has always been misunderstood, or, at least, been related in the wrong order. There is no movement apart from image, no image apart from movement. Spaces are no longer made of walls but of images. As Walter Benjamin noted, speed became the defining characteristic of the city and its inhabitants. There can be space in time, but not the other way round. This point of the dissolution of inside/outside– time and space merge into one. Or as Gaston Bachelard's metaphysical criticism of the separation of the interior and exterior suggests,

— "All really inhabited space bears the essence of the notion of home..the sheltered being gives perceptible limits to his shelter. He experiences the house in its reality and in its virtuality, by means of thought and dreams."

— Some architects work exclusively with the notion that the space must be conceived from the perspective of the moving body. Movement and architecture are opposing forces, although the instinctive pull towards creating timeless works of art also defies the concept we have of time. Do we need to rethink the relationship between our built spaces and the way we move in space within a framework of an elucidation of space-time that is fluid and temporal and as a continuum of optical and haptic experiences? Space is not a given but is created in time by the movement of bodies and sound waves.

— Merleau-Ponty refers to our sense of place as "the area of possible actions," the place where "there are things to be done," and from which "spatial meanings develop." According to him, small spaces can be depressing because these spaces are confining and prevent body movement. So the "lack of invitations to movement destroys my sense of being in place and at home in my world," he wrote.

— Currently, we can see the representation of some of these transformations, as digital artists and designers are developing interfaces that serve to extend the human body into its virtual environment, allowing the body to manipulate the surrounding space and its perception.

— The work of the architects Elizabeth Diller and Richard Scofidio, with their 2002 Blur Building in Neufchatel in Switzerland demonstrate this new way of interacting with the surroundings. A grid of computerized nozzles, supported on scaffolding and emitting puffs of mist, evoked an ethereal cloud. The exhibition is a suspended platform shrouded in a perpetual cloud of man-made fog. The so-called Blur Pavilion is visible many miles away as fog cloud. The building consists of a 60-by-100-by-20-meter metal construction that sprays innumerable tiny drops of lake water from 31,400 jets. The high-pressure spraying technology ensures that the fleeting sculpture will be visible in all weathers, rain or shine. Jets with tiny apertures force water with high pressure, forming little water droplets. The droplets are so small that most of them remain suspended in the air. If sufficient jets are installed in a specific volume, they saturate the air with moisture and create the effect of mist or, in this case, the effect known as the blur.

— Walking down the long ramp, visitors arrive on a large, open-air platform at the center of the fog mass, where the only sound to be heard is the white noise of pulsing water nozzles. Computers are adjusting everything from the strength of the spray according to the different climactic conditions of temperature, humidity, and wind speed and direction, as the fog mass changes from minute to minute.

— The Blur Building expands and produces long fog trails in high winds, rolls outward at cooler temperatures, and moves up or down depending on air temperatures. The really interesting aspect of this interactive project is that every visitor had to wear a coat with wireless technology embedded into it called "brain coats." Visitors' "brain coats" will react to each other, indicating either positive or negative affinity between visitors through color changes and sound. It created a powerful visual effect, instigating us to inquiry into the significance of the object as we experience the de-materialization of the object as we know it and the immersion of the body in affecting the outside environment. Truly, an interactive occurrence when outside and inside collapse into one, altering our perception of the world and ourselves.

ANONYMOUS

— Scientists should challenge themselves to utilize art within a scientific context. Cognitive neuroscientist Semir Zeki is an example of a scientist who collaborates with artists to advance his research. In his text "Art and the Brain" he uses examples of Cezanne's, Calder's, and Matisse's work to illustrate neurological discoveries supported through aesthetics:

— "...that we have learned enough about the visual brain in the past quarter of a century to begin to study the biological foundation of aesthetics. Aesthetics, like all other human activities, must obey the rules of the brain of whose activity it is a product..."

— http://www.neuroesthetics.org/research/pdf/Daedalus.pdf

— MFA Graduate Student – SVA

INGEBORG REICHLE

— In my research group we try to analyze the social and cultural implications of the images and pictures produced in the life sciences in collaboration with the Zuse Institute Berlin. This institution is a research institute for applied mathematics and computer science, and includes: Scientific Computing—Numerical Methods (Prof. Dr. Peter Deuflhard); Numerical Analysis and Modelling Visualization and Data Analysis; Scientific Computing—Discrete Methods (Prof. Dr. Martin

Grötschel); Optimization, Scientific Information Systems; Computer Science (Prof. Dr. Alexander Reinefeld); Computer Science Research.

— Our research interest is focused on the scientific production processes in which these images emerge and how some of these "beautiful" images are selected and used to communicate scientific research results to a public audience. But what do we see when we look at these images?

— With my contribution to this session, I would like to draw your attention to two very different research projects at the Academy of Sciences here in Berlin, two projects, which both try to analyze the social, cultural, and economical implications of bio- and neurosciences.

— Though we all work on the same floor and meet and talk quite often, it is very difficult even to understand the questions, the "frame" through which the social and cultural implications of the biosciences are looked at, because the appoaches are so different.

— The first project is the so-called *Gentechnologiebericht* (The Gene Technology Report). This research group aims to compile regular reports about the state of genetic engineering in Germany.[184] Many Academy members in Berlin are involved in this project, including Hans-Jörg Rheinberger:

— "Gene technologies, in other words recombinant DNA applications, have become well established in many fields of basic research as well as medical diagnostics, forensic science and the life science industry. It is extending its influence to ever new aspects of human life. These developments have led to fierce debates on principles and aroused emotions, particularly in Germany. What is desirable is an 'observatory' to observe and describe the advance of genetic engineering. There have been, and still are, a number of activities which describe the state of genetic engineering in Germany. There was a committee of inquiry, and there are information systems operated by industrial associations, government bodies, political and party institutions. Although numerous reports have already appeared with useful information on specific issues, they are not of the broad-based, interdisciplinary nature; they are not free from particular interests and do not draw on the continuous observation of the state and development of genetic engineering, as would be necessary to sustain an objective debate. A working structure consisting of an interdisciplinary research group with staff and an expert network is to set up with a view to establishing a centre of expertise on gene technology issues at the Berlin-Brandenburg Academy of Sciences and Humanities. Beyond the reports appearing at regular intervals, this will make it possible to respond swiftly to specific incidents, e.g., clones, accidents, unexpected problems and risks, in other words, issues at the focus of public interest. These issues will then be included in the report and documented and assessed in retrospect."

— The second project is called *Funktionen des Bewusstseins*. This project about biotechnology and consciousness is organized by philosophers and neuroscientists. The issues of these projects range from the question about consciousness in the age of biotechnology to the search for brain images before Kant. We have more than eight projects within this project, but I will draw your attention only to three of them, because these fit into our discussion: [185]

1) *The Molecular Person and Consciousness. A Study on Feeling and Thinking in the Age of Neuroscience, on the Basis of Immanuel Kant's Philosophical Anthropology (Fiorella Battaglia).*

— *Philosophy has approached the "problem" of the human being in various ways. Today the concern is to outline a holistic conception of the human through the joint efforts of both humanities and natural sciences—a conception that, faced with the new challenges of the present day, may also be able to give us a practical point of orientation for our actions.*

— *Taking as a point of departure Immanuel Kant's attempt to capture what it is to be human, this project will then describe an arc taking us to the anthropology of the present day, taking into account the current state of knowledge in neurophysiology and the philosophy of mind (Gerald M. Edelman, Giulio Tononi, Joseph LeDoux, Antonio Damasio, Christof Koch, Bernard J. Baars, David Chalmers, Daniel Dennett, Michele Di Francesco, and John Searle). Here the historical survey will place particular weight on that function of consciousness that can provisionally be characterized as I- or person-function that constitutes identity. The research in the neurosciences, in particular in their investigations into the metabolic, molecular, and functional activity of the human brain, presents an interesting and unique structure. This stems from that fact that in* ars medica *theoretical and practical dimensions are interwoven—in the person who, with his or her knowledge and actions, contributes to the health of another human creature. There is no theory of consciousness without practical reason. Thus the debate over consciousness overcomes the separation between nature and culture, in that brain research is the field that distinguishes itself in treating of elements of both these dimensions. Within the human brain, as Joseph LeDoux has shown, the genetic configuration and the individual experiences express themselves in the same language. It was Kant who opened the way for a dialogue between science and philosophy, between natural sciences and humanities, in arguing for the mutual interconnection of mind and body in a larger whole. His reflections on the scientific conceptions of the human being contain fundamentally insights into the conditions of knowledge in the natural sciences that are still current.*

2) *Modifying Persons: Self-Conception of Persons and Biotechnological Manipulation (Katja Crone).*
In the past years, great progress has been made in the field of neuroscience, expanding the spectrum of biomedical possibilities. Insights into cerebral functions have made the brain as the biological basis of human consciousness increasingly accessible. Thus neurosurgical, pharmacological, and genetic interventions impacting specific cerebral capacities have become increasingly feasible—capacities that correlate to specific functions of consciousness. Such interventions are likely to have repercussions for one's own self-understanding as an individual person, the way of conceiving one's own past, as well as one's possibilities for future actions, the way one interacts with one's social environment, and the reactions of others in return.

— *This prospect calls for an assessment of the possible outcomes of these new biomedical methods and thus gives rise to the main concern of this project, which aims at forming a philosophical concept of the person that systematically integrates key features of personhood. Of central importance will be the notion of a practical self, focusing on the concept of one's self-understanding as an agent and the ability to reflect and evaluate one's own practical decisions. The notion of personal identity over time will be articulated, founded primarily on a consideration of the specific nature of the first-person-perspective as well as the capacity for autobiographical reflection (narrative identity). Additionally, a concept of interpersonal recognition will be identified as a core issue of self-understanding and personal identity over time. The integrated model is meant to function as a conceptual tool allowing us to assess more precisely how personal functions of consciousness may alter due to biotechnological methods.*

3) *Representations of the Brain in the 16th Century: Models of Perception Between Physiology and Psychology. (Tanja Klemm, M.A.).*

— *Since the late 15th century, there has been increased interest in questions on the psychology and physiology of the human organism in regard to its perceptual faculties, not only in treatises of natural philosophy and theology but also in medical writings, encyclopedias, mnemotechnical texts and discussions of image theory.*

— *Psychology was at this time not treated as an independent discipline but as the philosophical study of the soul, placed by philosophers and scientists in the broader context of natural philosophy (and so also of medicine and biology). Not until 1575 did the German Humanist Feigius first use the term in the context of the Aristotelian writings* De anima *and* Parva naturalia. *Physiology, in turn, was seen as a system of interaction between body parts, organs, fluids and the so-called spiritus. This sensitive and subtle bodily substance floats from one organ to another, driving the natural (e.g., digestive), vital (e.g., affective) and animal (e.g., cognitive) functions of the body. Prior to William Harvey in 1628, this physiology of spiritus was the common conception of bodily organization: spiritus running through the body, being processed from the liver to the heart to the brain (from* spiritus naturalis *to* spiritus vitalis *to* spiritus animalis*) where it dynamizes mental, perceptive, and cognitive processes. In this conception of the body, the difference between mental and corporal processes is not categorical but gradual. Taking this as a point of departure, my project aims to investigate two fields: Firstly, it will focus on the understanding of the physiological implications of (visual) perception. Even though the subject here is the human brain with its imaginative, cognitive, and motive functions, I always consider it in relation to the whole body, its organs and functions. As the highest organ, though, the brain is by this time already seen as the seat of the soul and of the inner senses or virtues,* sensus communis, imaginatio/fantasia, vis aestimativa, vis cogitative, *and* memoria *(protected by the head, where most of the sensual phenomena arrive). One of the hypotheses here is that the conception of the psycho-physiological interplay between these senses is fundamental for an understanding of artistic concerns and pictorial theory of the time – and so also for questions concerning image production, the transfer of images, and their reception. A second field of research consists in the analysis of visualizations of the brain since the mid-15th century. Here, questions concerning the relation between two modes of visual representation of the brain/body are at the center of interest: the diagrammatic mode of representation (based on the traditional medieval psychology of faculties) on the one hand and, on the other, the representational mode based on sensual certainty (promoted, e.g., by early anatomists). One aim of these analyses is to contribute to a differentiated view on current tendencies in the neuro-sciences which put emphasis on visualizations and on the rhetoric of images. A historical perspective on these practices seeks to provide a sharper look on questions concerning the roles of images in the production of knowledge on the one hand and on the relation of brain visualizations conveyed by the media and the production of new knowledge of the brain and its functions on the other hand.*

FLORIAN DOMBOIS

— Bioscience seems to be an appropriate topic to show the relevance of different forms of research on the life sciences as "epistemic things" (Rheinberger). Even though scientific investigations have produced fascinating results, they do not seem to be able to catch up and explain adequately what life is. And many products of the life sciences are so irritating that alternative ideas are obviously needed. Artistic resarch is, in my opinion, that alternative.

VLADIMIR MIRONOV

— Dear all,

— One very popular contemporary form of bio-art is the tattoo.[186] In the former Soviet Union only criminals had tattoos. Now many Hollywood stars have tattoos, and tattoos are very popular forms of body art in the younger generation. There are several journals especially devoted only to tattoo art. Thus, whether we like it or not, tattoo is an integral part of modern art.

— Tattoo, as any fashion, is a temporal event in relation to historical time and place.

— One of the unexpected applications of the modern genetic revolution is the technological possibility to create removable tattoos using conditionally induced pro-apoptotic genes. Apoptosis is a programmed cell death. If we will create a library of living cell aggregates that express different colors employing fluorescently or nonfluorescently colored proteins and if these cells also express conditionally induced pro-apoptotic genes, then tattoos can be removed by the consumption of tablets of antibiotic tablets. Tetracyclin will induce cell death in these genetically designed colored cells.

— There are several advantages of this new living removable tattoo:

— People can re-use their body surface for tattoo body art; if people do not like their tattoos (for whatever reason), they can be removed without any surgical manipulation; it is possible to create an invisible tattoo that could be seen only by particular lengths of light waves.

— Sometimes new technologies create unusual and original unexpected applications.

— By the way, the modern tattoo machine was invented by Thomas Alva Edison *[fig. 61]* in order to make paper copies of texts by using mechanical pinching. [187]

— My questions are:

— Is there any social demand for such a type of tattoo?

— Can living removable tattoos be considered as modern living bio-art?

— Is a living removable tattoo a socially acceptable and socially responsible practical application of the genetic revolution?

EUGENE THACKER

— A friend of mine—who wrote freelance for the biotech industry journals—once said that the "killer app" for the biotech industry would not come from medicine but from an unrelated or spinoff industry—e.g., food, detergent, household cleaners, toys, fashion, computers, even pets. He also said that this would serve to "cushion" the introduction of biotech into the everyday. I assume he was thinking about the prior example of computers—from military mainframes, to IBM business machines, to our "personal" laptops, pods, and berries.

ORON CATTS

— The use of biomedical technologies for nonmedical ends is one of the criticisms that is being waved against artists who are working in this field.

— I remember that Vacanti once said that the goal of tissue engineers is either to get the Nobel Prize for medicine or to find a way to regrow hair.[188]

— When we presented one of our pieces as a poster in an international tissue engineering conference, we recieved a hard time for degrading this technology to something as frivolous as art, although some posters next to us dealt with penis extensions and other cosmetic enhancements. As the technology of producing 3-D organs using tissue engineering proves to be as elusive as ever, this technology is starting to be employed for nonmedical ends. Body enhancements is an obvious way to try and manufacture consumer needs for this new technology. Others are the development of in vitro meat and leather industries (that we had something to do with), as well as military applications.

TROY DUSTER

— The relationship between bioscience and society is contingent on the particular society in which the bioscience is translated into practice and policy. Here are two easy examples: genetic screening and testing of a population in a society with universal health care is much less of a contentious issue than gene screening and testing in a society in which insurance companies can root out and exclude those with asymptomatic "genetic" conditions—from late onset disorders such as Huntington's to likely risks of those with BRCA [breast cancer] genes. Or second, imagine the different rate of acceptance by Ashkenazic Jews for Tay-Sachs screening done by Michael Kaback in the last two decades of the twentieth century (a near-complete success story) versus a similar hypothetical program in the Third Reich.

— Two kinds of national DNA databases are in our future. The Portuguese government has already approved "everybody in the DNA bank" legislation in early 2005. The British now have over four million samples from their population in a databank, with an explicit goal of getting a quarter of the population in by 2010. One such database is designed for forensic purposes. As in my brief note above, acceptance, compliance, or strong resistance will depend upon the kind of society. The British were primed and ready for expanding DNA databanks from thirty years of acceptance of what in the US would be resisted (ACLU) as draconian counterterrorist activities spinning out of the northern Ireland conflict (cameras, cameras, cameras everywhere…DNA dragnets were pioneered in the UK, etc.). But a health-inspired national DNA databank is also a likely development in some societies; again, note the conditions from the first entry above.

— In both China and India, there is the beginning realization that sex-selection practices now assisted by technological breakthroughs (from expensive amniocentesis all the way to cheap ultrasound) are dramatically distorting male and female relations, potential mating opportunities, and will have a profound effect upon social and economic policies that few could have imagined. The book *Bare Branches: The Security Implications of Asia's Surplus Male Population* [189] is a harbinger as ominous as global warming scenarios (e.g., for polar bear extinction) about the myriad of possible spinoffs from sex selection. The Chinese government is now trying to initiate some "correctives" and incentives, but thirty years of female infanticide via selective abortion is not so easily "corrected"—thus, for contemporary China—and for the next decades, lots of "bare branches."

[public blog]

NATASHA VITA-MORE

— Insightful article. My thoughts are that people must become aware of the pros and cons of DNA transparency within the jurisdiction of their respective communities. There are trade-offs in life and we need to critically assess what the trade-offs are for each one of us, knowing the condition—map of our physiology as well as our cognitive properties—of our DNA. Not knowing the condition of our bodies is like driving down a highway at ninety miles an hour blindfolded! In the future, we very well may look back at these times wondering why humanity worried about the transparency of their DNA as their lives pass them by. I would rather know what disease I have a propensity for and deal with it than hide my DNA so no one could charge me an extra sixty dollars a year for the exposed cancer gene. If we know that fifty percent of humanity will get Alzheimer's by the time we are sixty-five, then I would venture to say everyone would be willing to donate a percentage of their income toward curing that particular disease.

ANDREW CARNIE

— Understanding of genetic diseases and their transmission from one generation to another became the subject of the work that I completed for the Mendel Museum in Brno in the Czech Republic called *Things Happen.*[190] I collaborated with Bernadette Modell, a population geneticist, Emeritus Professor of Community Genetics, UCL [University College London] Centre for Health Informatics and Multiprofessional Education, London *[figs. 62, 63].*

— In some cases, the knowledge held by families and social groups was used to counteract such effects and was significantly touching. A small understanding of this field gained from making this work made me feel very strongly that we are all interrelated. It was a powerful antidote to Margaret Thatcher's (I can hardly bring myself to type her name) conviction that, "They're casting their problem on society. And, you know, there is no such thing as society."[191] I would hope that contact with knowledge of genes and genetics through art and science would help us realize our interconnectivity and interdependence.

RICHARD TWINE

— I don't think this has been mentioned yet, but it occurs to me that we should be a little cautious about valorizing the lab as the site of scientific investigation when talking about art-sci crossovers. During my research into animal geneticists and genomics scientists, it became very apparent that less and less work, and in some cases no work, is carried out in the lab as we might traditionally conceptualize. Many of us will be familiar with the idea of life being converted into code or information.

— Well, I saw this in practice—the majority of work is now done in the office in front of a computer screen, doing comparative work on various cross-species databases of sequenced code. Livestock animal scientists draw upon the sequenced human and mouse genomes in order to learn— through homology—about quantitative trait loci of economic interest in their respective species, be that pig or whatever. Obviously lab work still happens, but there is a degree of stratification. It's more likely to be the grad students or post-docs doing the lab work. This may be something to think about or of use to someone.

ORON CATTS

— The point that is important to note is what it really means with regards to these types of technologies, by employing which reductionist abstraction, the victim is hidden from view. The notion of the technologically meditated victimless utopia that our last few projects ironically dealt with is indicative of the distance that Western capitalism generates between the perpetrators and their victims (see the media coverage of war in the

US). This type of distance can be seen as a device to eliminate the need for contemplating the ethical issues so not to hinder "progress." The visceral experience that much of the wet bio-art produces might be seen as a way to confront this notion.

SUZANNE ANKER

— Novel ways of thinking and doing can be socially, economically, and politically beneficial. The wonder of invention is a creative springboard allowing the more adventurous to remix the given. Negotiating territorial discourses and practices requires tenaciousness, and makes significant learning demands on the border-crossers. Mironov's suggestion about living, removable tattoos as possibly being a form of bioart, is something to think about. Certainly within the traditions of body and performance art, possibilities abound. Many artists have used their own bodies as malleable sites of sculptural form. From Joseph Beuys and Bruce Nauman to Hannah Wilke and ORLAN, performance art is very much alive within the cultural establishment of the art world. In an earlier post, ORLAN talks about a prospective collaboration with Oron Catts. Mironov's idea about removable tattoos may, in fact, one day cross the art-sci divide as well.

— Invention can also be looked upon as incorporating an element of chance. There are many historical examples in this regard, the discovery of penicillin being one. For the artist, recognizing nuance and possibility in the process of making is part of artistic creation. Another aspect furthering collaboration and creative process is the random and not so random meeting of various practitioners at cocktail parties, lectures, art exhibits, and the like. The scientific laboratory and artist's studio are generally off-limits to a public audience. Perhaps shared laboratory/studio visits could be arranged?

— I am very intrigued to learn about Thomas Edison's tattoo machine. In some ways it can be compared to Paul Winchell's prototype for an artificial heart. The comparison is not one of function, but one of migration between the worlds of art, science, and entertainment. Most Americans (of a certain age) are familiar with Paul Winchell and Jerry Mahoney, as a ventriloquism duo performing on early television. Paul Winchell being sentient, Jerry Mahoney manufactured as a puppet. However, what is not well known is that Winchell was also an inventor. He studied acupuncture, was engaged in medical hypnotism, and had a close relationship with Dr. Henry Heimlich. In consultation with Heimlich, Paul Winchell designed the first the prototype of what we now know as the artificial heart.

[public blog]

GABRIEL HARP

— I want to reflect on some aspects of the conversations from the symposium on visual culture and bioscience and try to reframe the discussion with notes for myself.

— A focal point for the discussion has been to discuss the roles artists play in laboratory settings. A good deal of discussion and effort has been directed over the years towards promoting opportunities and understanding the roles that artists perform alongside and/or despite scientists. Suzanne Anker also introduced the notion of "art as invention"—the migration of social spaces in a recent thread dealing with the social and cultural implications of bioscience.

— While these are complex topics full of social and methodological implications, I sense a general way of framing the conversation that, I think, warrants some consideration—if only because it might help draw others into the conversation.

— Whatever the engagement of art and biology, be it through individuals, groups, and/or at the level of disciplines, there are certain instances of value created by these multifaceted interactions. Many of these have been discussed. However, these are not characteristic of only art and bioscience, but rather they seem to exist at all levels of entrepreneurship. Entrepreneurship is the practice of sensing opportunities and starting new organizations (of concepts even!) to seize on those opportunities.

— I like Suzanne's sentiment—though I have to disagree about the role that chance plays in the interaction. Certainly there is an element of chance, but invention is much more an import/export process: "Negotiating territorial discourses, recognizing nuance and possibility, meeting of various practitioners." From my point of view, the majority of what counts as interdisciplinary/transdisciplinary practice involves translation, mediation, organization, and consultation. If you are involved in an actively engaged import/export process of exchanging ideas, materials, methods...whatever...you generally increase the odds that chance will fall in your favor. The implications, especially for teaching and learning, are that the import/export process can be learned and developed, while chance cannot.

— Would you agree that a characteristic of artists is

their ability to sense gaps in culture or social structures? Would you also agree that artists are able to bridge these gaps because they maintain affiliations and learning strategies that involve and leverage the concerns of multiple groups?

— I see at least two ways of looking at this:

— Behaviorally—that is, what are the behaviors of artists and scientists that seek to make these kinds of bridges and border-crossings? A question to ask here is, "How?"

— Functionally—that is, what are the benefits that these behaviors confer both for the individuals involved and for their "other" constituencies? A question to ask here is, "For Whom?"

— From the behavior standpoint, this idea of entrepreneurship is critical. I admire the symposium participants' overriding desire to identify new areas of investigation, their resourcefulness in building a practice and implementing projects, and their dedication—despite conventional "wisdom."

— Woven through this discussion have been themes of

— – creating common ground,

— – creating new holistic understanding, and

— – resolving differences between disciplines through the development of a metaphor—visual or otherwise.

— These themes, not coincidentally, characterize Wolfe and Haynes's criteria for interdisciplinary synthesis (at least in terms of writing).[192]

— To what degree are the behaviors, practices, and methods of the artists and scientists involved in this discussion and elsewhere indicative of brokering–such that each is translating, applying metaphors, creating common ground, and resolving differences between groups? This brokering, I think, is a fundamental activity and the research seems to indicate that it can be learned.

— For much of my argument about brokering, I am drawing from Ronald Burt's work on structural holes and network entrepreneurship. Burt recently presented to NESTA [National Earth Science Teachers Association] on the topic of innovation.

— The domain knowledge and landscape is also critical. The fact that so little is known about biology, and for that matter about art, necessarily creates opportunities. These are amplified by the apparent and real methodological differences between biology and art practices. Thus, one explanation for why this is such a rich area of inquiry lies in the social, cultural, and ethical implications of bioscience—confounded by the social and cultural "holes" created when we try to align the constituencies and interests of biology and art.

SUZANNE ANKER

— I offer the following in response to your astute reflections. Your point concerning "multifaceted interactions" does, in fact, apply to a range of practices going beyond the scope of art and bioscience. Leo Steinberg in his 1953 essay "Art and Science: Do They Need to Be Yoked?" addresses some of your questions. He sees the art-sci connection as only one of the possible pairings of creative thought. He so elegantly asks, "Don't any two disciplines offer mutual analogies to a rhetorical imagination?"

— With regard to "import/export" process in knowledge production, chance remains a variable x-factor based on many other surrounding socio-economic and political backdrops. Certain practices: laboratory, aesthetic, or otherwise are also influenced by outside variables of funding, cultural readiness, and media-driven targeting. So it is not the practice itself that is immune to the timing of its reception.

— Let us consider this example. I design a cooking contest in which cooktops, utensils, frying pans, butter, and eggs are given to twelve contestants. Each ingredient has the same expiration date and was purchased at the same store at the same time. The frying pans, as well, are of equal value. Each contestant is then asked to prepare "eggs over easy." A panel of three judges are asked to award prizes based on taste, presentation, and alike of the food. How can we assess the differences in quality of the results? Doesn't the incorporation of chance through performative skills and subtle cues affect the outcome?

— Your points about translating, applying metaphor, creating common ground between disciplines is well taken. In fact, I would like to see a small think tank take place in which presentations, discussions, and performative projects would help to clarify differing discipline-based interpretations. Since specific expertise generates specific epistemological results, assumptions can be reframed in terms of what means what to whom. Since we are dealing with specialized knowledge bases, there is much to learn.

— In lieu of being able to participate in the debate, through sheer pressure of time, I'm sending my latest *Nature* column, which is at least germane. It is not yet published and has not been through the editorial process.

— Sarah Jacobs has been keeping me informed about her work for a number of years.

— We met for the first time when I was preparing this essay. I think she will come to be regarded as one of the major artists working in the field.

— Martin Kemp, "Science in Culture: Gene Expression."[193]

— *How is any artist to confront the excruciating complexity of the human genome – or any genome for that matter? There are just too many CGATs [Comparative Genome Analysis Tool]. It is possible to make some general artistic "statements" about the project, about its implications, and about genetic engineering. It is all too easy to sink to the level of the "Frankenstein Food" headline in the* Daily Mail *on 13 February 1999.*

— *But Sarah Jacobs shows that the complexity can be tackled head on and turned to brilliant aesthetic account. She has a record of working with the blank poetics of modern scientific discourse, with its studied eschewing of stylishness or personal expression. Her ninety-page book,* Deciphering Human Chromosome 16: We Report Here *[fig. 64] is studded throughout with phrases from the original 2004 article, "The sequence and analysis of duplication-rich human chromosome 16" (*Nature *432). "We report here" is one of these, together with "We observed" (of course),"Here we describe," "We constructed," "We adopted a strategy," "We then eliminated," "Finally we identified," and so on. Isolated, the phrases that are so much part of scientific normality assume the quality of an incantation.*

— *Following the* Nature *article, Jacobs googled such terms as "human chromosome 16," "chromosome 16 book", and "chromosome 16 expression." She even searched for odd combinations, such as "chromosome 16 Saddam Hussein." (Yes, it really does produce results.) She sifted out around 250 website links on the basis of what appeared intellectually or intuitively interesting and "looked good." The e-book proceeds through simple pages of the incantatory phrases interspersed with colored lowercase overprinting of the site links with fragments of their texts and numbers from the original article in large capitals, such as the ".....NINE PERCENT / EIGHT HUNDRED / AND EIGHTY. ONE THOUSAND / SIX HUNDRED AND SEVENTY / NINTEEN / THREE HUNDRED AND / FORTY-ONE / 3," on the illustrated page.*

— *The result is a doggedly accumulated "report" on the incredibly rapid Internet diffusion of the knowledge in standard and bizarre forms. The contents are, however, subject to constant mutation. Every six months Jacobs took screen shots to document the changes.*

— *To accompany the* Report, *she has now issued an* Index *as a print-on-demand book, heavy in its fixed form of 552 pages. Against the rat-a-tat background of the CGAT permutations, the accumulated numbers are remorselessly spelt out, up to "Sixteen million five hundred and forty-one thousand and nine hundred," still short of ninety- million-plus noted in the article. They are accompanied by enigmatic fragments from the websites.*

— *Given the vagaries of the production process, each* Index *assumes an individual character. The CGATs on every left-hand page are bled to the very borders, and their visible expression along the unbound edge of the closed book varies unpredictably as the result of minute variations in the trimming process.*

— *The* Report *and the* Index *are strange, difficult, perplexing, suggestive and strangely beautiful - and awesome in their numerical persistence. Jacobs has created something that is very directly drawn from the science and its diffusion, using the tools of a rabid bibliographer-cum-classifier. Yet the result subverts the science in the direction of chaos and cacophany. The effect is analogous the way that the extraordinary particularity of each individual person seems to confound the overwhelming similarity of our genetic constitutions.*[194]

— *At least this is one interpretation that I can give it. There are others. Jacobs is, I suspect, resisting any closed or dominant reading. Therein lies the difference between the original article and Jacobs's visual play. The scientific exposition states that "WE FOUND" with as little lattitude for alternative readings as possible. Jacobs provides a field for interpretative flexibility that triggers thoughts and insights of an unexpected nature – unexpected even to the author herself.*

SUZANNE ANKER

— Several panelists have had professional training and/or careers that span the visual arts and the sciences. Florian Dombois comments on the ways in which content from one discipline can be transfigured and expanded into components of another. Please talk about the ways in which symbolic forms intersect with your work in the sciences. To what extent do you differentiate your roles in each discipline? What are some of the ways you can describe art practice as an epistemic system?

RICHARD TWINE

— As a sociologist interested in the social and ethical aspects of animal biotechnologies, a central issue is to investigate the extent to which the biosciences are affecting a shift in our relations to nature, and the very meanings of "nature." I take on board Oron's point that artists, sociologists, and cultural critics can certainly become embroiled in the mythical hyping of bioscience, and so consequently we should be cautious in overstating our claims. Interestingly, since we are talking about highly commercialized domains of science, scientists may be encouraged to adopt a rather performative stance to prospective venture capital audiences. We can witness this right now in the area of using cloning technologies for animal agriculture, which (setting aside ethical questions and focusing on the technical possibility) is rather "gung ho" in the USA and treated with much more scepticism in the EU.

— We're in the early stages, but arguably changes are afoot, hyped or otherwise. Marker Assisted Selection or Gene Assisted Selection, which are molecular techniques of refined selective breeding using genomic information (not to be confused with genetic modification - GM), are developing with some already in use, and pockets of research for the use of GM and cloning in animal agriculture are also ongoing. Xenotransplantation has hit a wall, but research continues, and biopharma research, although rather slow in progress, is also ongoing. Several of these techniques encourage a convergence between medical and agricultural domains, which is novel in its hybridizing effect, if not its history of knowledge exchange. Obviously plant agriculture is further "ahead" in these respects.

— The ethical import in these transformative biopractices could revolve around the question of what the human becomes when it transitions into a designer of life. The biosciences might equally be drawn upon to underline cross-species solidarity or indeed a definitive humiliation of a nature still represented as abject and nonhuman. In terms of animal ethics specifically, the vast majority of GM work is in medical research, a historical paradigm that is very difficult to extract ourselves from. Nevertheless, our values towards other animals are highly ambiguous, with this ambiguity probably being a latent cultural resource for change. It's arguably the dark little marginalized truth pertaining to our cultural practices of animal consumption that is more ethically interesting. Unless our thriving posthumanist vegan and vegetarian friends are grabbing some animal protein on the sly, it does in fact appear that our animal consumption is a psychological/cultural/geographical need rather than a physiological one.

— We might then want to think more on why art-sci interactions could be seen to be privileging certain forms of "sexy" science. An artist working with a team of animal welfare or behavioral scientists or even nutritionists might come up with a different take on the issues of hubris and anthropocentrism than one who goes for animal transgenics, as interesting and as valid as that is. Perhaps, as implied by one or two earlier posts, sciences such as tissue engineering will have a considerable impact on human/animal relations if engineered meat really does happen, or at least a proliferation of novel protein foods.[195]

— It would be good to hear more from the artists here who use animals in their work—whether or not they think their work is raising any particular ethical questions or questionings of bioscience.

MICHAEL SAPPOL

— This may be stating the obvious, but I want to offer an abstract, but I hope useful, response to the virtual symposium as a whole. Oron Catts's post about the unfriendly reception of his poster at an international tissue engineering conference raises the issue of the ambivalent relationship between artists and scientists, and their work. Without invoking the now-tired "two cultures" argument, it seems evident to me from many of the posts in this symposium that there are structural, and more than structural, similarities between artists and scientists: both make artifacts that are connected to truth claims; both do their work in sequestered spaces (the laboratory, the classroom, the studio, the gallery, the journal); both derive authority and legitimacy, cultural privilege, from claims to be connected to progressive lineages of accomplishment, with chronologies of "landmarks" and "breakthroughs"; both have a canon of performance based on the notion of "genius"; both increasingly use a specialized technical vocabulary that excludes the larger public; both have knowledge communities that assess, comment on, and validate their claims (and both share a knowledge community, the group of people who do sociological/cultural theory/critique of art and science); and both have transformative effects on a wider public, on society and culture, on everyday life, on the embodied self.

— Although some of the posts in this symposium have stressed the ways in which artists and scientists collude, and have colluded over the centuries, it seems to me that there is a competition for authority, an abiding tension. Today, the cultural prestige of science is very high, hegemonic; for a long time science has trumped art. Artists have responded in part by adopting the language of "experimentalism" and the "laboratory," first in music, then in conceptual and visual art; by working in new technologies (video, digital imaging, tissue engineering, recombinant DNA); by taking up residence in the institutions of science; by adopting, at least in part, the specialized technical vocabulary of science; by making "epistemic objects" that mimic or parody or "destabilize" the objects of science; and by taking on an oracular, prophetic role that offers a critique of science, especially biotechnology. In other words, artists have responded by making themselves into scientists and philosophers of science, who make far-reaching truth claims about science and technology: about the way science works; about the power relations that are built into science's "epistemic objects"; about the mystification, reification, and/or naturalization effects of scientific productions; about science's embedded ideologies and aesthetics; about its cognitive, and much more than cognitive, effects on us and the world around us.

SUZANNE ANKER

— Having recently re-read C.P. Snow's *The Two Cultures and the Scientific Revolution*[196] with my M.F.A. graduate students in SVA's Writing and Art Criticism Program, we did, in fact, find some points that still have currency. Although couched in existential terms, Snow distinguishes "the individual experience and the social experience, between the individual condition of man and his social condition." Thus for Snow, who had one foot in each camp, he talks about scientists thus: "Most of the scientists I have known well have felt just as deeply as the nonscientists I have known well— that the individual condition in each of us is tragic. Each of us is alone: sometimes we escape from solitariness, through love or affection or perhaps creative moments, but those triumphs of life are pools of light we make for ourselves while the edge of the road is black: each of us dies alone."

— Your impressive and comprehensive listing of "structural similarities" between art and science is, in fact, well taken. However, I would like to argue with your notion of "artists making themselves into scientists and philosophers of science." I have made references during the symposium to the fact that, at the present time, art practice has devolved into an entertainment industry. Operating as an unregulated insider trading brokerage network, it advances its platforms and cultural consensus through a checkbook. From the art fair to the international biennial, goods for sale (or tourism bucks) are only outflanked by the perception that the "expression" of art is synomous with human rights and political freedom. The enormous interest in objects and markets may, in fact, dilute the actual practice of art. If one conceives of the art world as a microcosm of the real world (and this may or may not be true in science), then the commodity trumps the iconic, linguistic, philosophic, or other charges that have historically been within the provenance of modern art. And it is here that we can consider the changing ways in which communities and social complexes form. As global markets produce more and more "stuff," objects, as stated by Karin Knorr Cetina, replace personal relationships, substituting "things" for sentient intimacy.

— If artists are migrating towards alternative discourses, it may, in fact, be that they do not agree with the art world's self-proclaimed agenda. In this sense, artists are still doing what they have always done, particularly in regard to the historical avant-garde. The claim that art has value exceeding its material costs has created a coterie of art historians, historians, curators, documentarians, and others, as they interpret, catalogue, and protect what has been deemed valuable. So, in some ways, art functions as a databank or archive of the changes over time of our ideas about what constitutes "world-making." Hence, this current round of artists, not content to endorse the slogan "dumb like a painter," wish at least to have the "creative moments," the "pools of light" that C.P. Snow talks about. And it is to this ambitious undertaking that I give my respect.

EUGENE THACKER

— Hi all,

— I have to admit that I concur with many of Michael's statements here, though I remain interested in innovations in both and across both areas. Personally I'm not convinced that we are anywhere near being "beyond" the two cultures (let alone in a "third culture," as John Brockman suggests).[197]

— There's a conceptual issue involved, in that we often "begin" discussion presuming the categories of "artist" and "scientist," if only for heuristic reasons. That's fine; we all do it, whatever. But there's an important discussion to be had about the function of those categories in making possible a discussion at all. The artist-scientist dichotomy often comes to metonymically stand in for the "two cultures," whereas they are far from being identical. I'm not sure a philosopher would be on the side of the artist, or a programmer on the side of the scientist, etc.

— But I'm always reminded of the way that the various institutional sites relentlessly—and invisibility—interpellate us into these positions as well. I would include here the very literal spaces of the classroom, the art studio, the media production "lab," the science lab, the conference/convention/symposium, and online variants of these. There's no reason to think that Peter Burger's[198] critique of the "institution of art" doesn't apply to other fields as well.

— This conceptual issue, and its articulation within institutional sites, leads us to think about the refractory effect that many of our basic concepts have. For instance "creation" may mean something very different for one artist and another, let alone between those artists and a scientist. How different can the concepts become while still remaining the same concept? (The concept of "purpose" is another one; Kant famously posited a weird "purposiveness without a purpose" in the asthetic.)

— A while back I tried to—in a somewhat provocational manner —sort out some of these problems in a short article that started by flaming Jeremy Rifkin and then talking about bio-art. I quote from a part of it here:

— *Bioart usually benefits the artists more than the scientist collaborators. While there are a great many examples of scientists collaborating with artists on projects, there are a few asymmetries worth noting. First, the work itself is usually shown in an art context. Second, if publication occurs, it is more likely to be in an art journal than a scientific one. Third, when instances of professional recognition arise (e.g., tenure and promotion), the artist gets recognition, while the scientist often does not. Fourth, artists and scientists work with very different funding budgets. Very different.*

— *The context for bioart is often the site of the gallery. This may not be problematic in itself, but when bioart claims to be speaking about biotech in terms of education and public awareness, then we have to wonder about the site of this engagement. The art gallery is itself a specialized site, quite alienating for many people. How can art claim to reach a public about science, when it still has not resolved its inability to reach a public about art?*

— *In bioart, "gee-whiz" science often overwhelms critical engagement. That is, bioart often eschews ethical considerations in favor of technical ones. Anyone will admit that learning how to work the automatic sequencing machine is cool, but it is worthwhile to reflect on it a little. The old question, can I do this versus should I do this, is worth reconsidering in the context of bioart practices as art practices.*

— *Bioart can sometimes become PR for the biotech industry. In some cases the aestheticization in bioart can feed into the "rhetoric of wonder" abundant in popular discussions of the genetic understanding of life. It is fascinating that your DNA stretched out is five feet long (or whatever it is). And?*

— *But not all bioart is formalist. In fact, a number of artists enjoy and cultivate the "outsider-artist" persona, which indicates that bioart may be attempting to fashion itself as the new avant-garde (oh no, not again!). By pitching itself as transgressive, bioart risks replaying the tired narrative of mainstream recuperation. Except that recuperation will this time be activated by government research institutions and biotech companies with programs titled "a celebration of art and science." (Might we someday see artists as spokespeople for pharmaceutical companies?)...*[199]

[public blog]

ANONYMOUS

— Eugene,

— Bio-art is an indication that science is an area where some of the most innovative, consequential, and exciting investigations are happening today.

It is natural for artists to migrate towards this territory since that is in keeping with the ways that modernism's concept of innovation continues to unfold. Does it matter who benefits more, the artist, the community, or the scientist? We are not talking about "Big Science" in this frame of reference where millions of dollars are at stake. Nor are artists reaping significant financial rewards, since their hybrid practices are outside of mainstream parlance.

— If the end result is more in keeping with knowledge production, everyone benefits. When speaking in terms of education and public awareness, the transparency of opaque issues provided by greater dialogue can only aid in discovering what the issues really are.

— Since 2003, when the Rifkin article was published in the Guardian, much has changed with regard to bio-art.

— MFA Graduate Student - SVA

VLADIMIR MIRONOV

— Popularization of biotechnology through bio-art is an excellent and reasonable idea. Scientific conferences are probably not the best place for bio-artists to present their works. Criteria and goals in science and bio-art are different. However, science and art museums are probably the most appropriate places for bio-art. Bio-art can help to attract a new generation of scientists to the bio-industry. Bio-art can also create a much broader public appeal and impact, especially in the discussion of potential social aspects of the adaptation of new biotechnologies.

— Concerning the suggestion, "Might we someday see artists as spokespeople for pharmaceutical companies," any new technology in the drug or medical device industry usually needs and uses professional visual presentation and even animations. Thus, it is safe to say that in certain aspects, artists are already working as spokespeople for the biotech industry. One can predict that the contributions of bio-artists to the bio-industry will continue to increase. However, the interaction between bio-art and biotechnology, their mutual enhancement and mutual benefit, is definitely beyond just pragmatic considerations, and it deserves special systematical analysis.

[public blog]

BRAD P.

— As an outsider to the biotechnology world, I had the privilege of discussing this topic just a few evenings ago with one of your contemporaries. I agree that, presently, the artists' contribution is primarily as a "sales tool," to attract promising young minds into this field.

— Looking forward, however, it is not inconceivable to imagine the artists' tools being advanced enough (consider the advancement of the artists' tools over the past decade) to take research data and develop conclusions (okay...call them predictions) through the use of these tools. Certainly, there will be learning curves associated with the analysis, but that exists with most analytical tools.

— The marriage of these two professions seems synergistic, as both thrive on innovation and creativity. I enjoy peering in on this strange, vast world of scientific and cultural exploration. Thanks!

MICHAEL SAPPOL

— Suzanne, I like your response to my "two cultures" posting. But I think you misunderstood what I was getting at with my statement that artists are "making themselves into scientists and philosophers of science" (because I was operating in the ironic mode). I was thinking about performative aspects of contemporary art. I was arguing that artists play-act the discourse, tropes, productions, and characteristic (or stereotypical) performances of science. This work, like other contemporary art, participates in the creation of marketable personae and trendy commodities. But in the best cases it also performs a valuable critical function—in collaboration with the apparatus of art historians, documentarians, etc.—mounts a critique of science, art, governmentality, markets, and lots of other things that need thoughtful explication and criticism.

ANDREW CARNIE

— I would agree with what Suzanne is saying; there is a drive to move away from some of the maddeningly commercial-driven aspects of the art fair and art market. Certainly some sanity can be found in the discourse and images displayed within this symposium. The web offers other ways of displaying material and new technologies, new ways of presenting material. This means artists can subvert the traditional dominant cultural forces and make their own production and consumption zones. Of course, they then have to accept the limitations that come with such a domain.

— With regard to the two cultures and notions of dominance of one over the other or considerations as to whether they are coming together as has been suggested, certainly I think mixing the two cultures up is healthy. Having deeper understandings of every field of life is important. I think artists should be very happy to talk and work within any field of study.

— Maybe there will be a new phase and a new fusion of science and artwork, and this will produce new types of production that will supersede our traditional separations. However, I would venture to stand by the fact that, ultimately, I think the scientist and the artist have to stand back from any excursion into the other's fields in order to engage in good science and art from the heart of their own practices, respectively. I think that artists need to have content and they need to know it well. This content can be in a scientific field. Specialists will have to reign, ultimately returning to what they know best and successful multidisciplinary experts will be very rare. I have written this, now do I believe it? Give me a few days.

— As a practitioner, I don't feel any particular jealousy towards scientists and what they discover. I am amazed by what they do produce, and I am amazed by what I have produced and what other artists have produced over the centuries. I have talked in conferences amongst scientists, having been overwhelmed by their discoveries and ideas, only to find at the end that they have been very impressed by what I have shown them in terms of my practices. The images and metaphors produced by artists are obviously powerful things.

— Strangely, maybe it is often the art that seems to linger longer in memory from any one generation or from any past century down the years. Is this a controversial thing to say? Is it only true of the past, only true of me? Will this current generation and future generations remember only science from the period we are in now? With science so dominant, are there cultural icons currently being produced that will pass on into the collective image bank that future generations might carry?

— I can see an image of the Dutch anatomists at work in my head from the seventeenth century, but can remember little of the science of that time *[fig. 65]*. Likewise, I can conjure up the images produced by many artists from the Renaissance, but remember little of the fascinating discoveries made in Florence. Ultimately the products of science are rules and formulae about the world that are ever being advanced. "Significant" art deals with particular human "subjective" truths, and these strangely become critically and objectively "true" generation after generation. Art is about sounds, pictures, poetry, and prose, and they seem to linger for longer. The science writing that Richard Wingate talks of in his recent post is, as he says, "paired down to exclude all metaphor, and has no place to resonate."

— The visual and audio world we know has existed for humans for much longer than that of the spoken and written word. We have always seen and heard, but we have only begun to talk and write more recently, with the first of the earliest known hieroglyphic inscriptions around 3200 BC. Homo Habilis was looking and hearing in Africa some two and a half million years ago.

— The picture and metaphor may have a deeper resonance with older areas of the brain. I rather like Richard Gregory's visual experiment with the hollow mask, spinning slowly. The nose always appears to stick out, even when the reverse should be true. Our brains overrule our eyes, to say "that the nose always protrudes"; perception acts over sensation.[200]

SUZANNE ANKER

— As we embark on our final, last words segment,[201] please feel welcome to preview any forthcoming projects you are engaged in. Additionally, please feel free to express your thoughts in regard to any of the ideas posted here.

BERGIT ARENDS & SABINE FLACH

— Dear all, Sorry to have been a little absent from this symposium. Sabine Flach and I have been at a conference abroad and are actually now sitting in front of the computer in London before dashing to the Big Apple. So, thank you all very much for the interesting debates.

— Sabine Flach is Head of Research for Arts and Sciences at the Centre for Literature and Cultural Studies in Berlin and has become known for her lecture series *"Wissenskuenste"* ("The Arts of Knowledge and the Knowledge of the Arts"), in cooperation with the Museum of Contemporary Arts Hamburger Bahnhof in Berlin. Sabine is a historian of art.

— Bergit Arends is now Curator of Contemporary Arts at the Natural History Museum, London. Beforehand, she headed the arts and science funding program supported by the Sciart Consortium and later by the Wellcome Trust.

— At the time of the sci-art program, I was lucky to be able to fund a collaborative project by artist Jacqueline Donachie and scientist Darren Monckton entitled *Tomorrow Belongs to Me.* The work was shown recently at the Hunterian in Glasgow, and a publication followed. The project deals with the cultural and ethical prejudices and the undoing of these prejudices in the study of "anticipation," a genetic condition that determines Huntington's and Myotonic Dystrophy as well as Fragile X syndrome.[202]

— Currently Sabine and I are developing an exhibition on the basis of recent advances in neurosciences and scientific research into emotions and its interrelations with the arts and literature. This research doesn't directly spring out of the work in the life sciences, but is an example of cultural coding of the body and how this is read through the sciences and, in this case, the boundaries of the neurosciences, i.e., where the sciences seek recourse to artistic production to further their research.

— Sorry, all the best, more from NY. The cab is outside.

CARL DJERASSI

— The program introduction of my play *Taboos,*[203] which premiered in London in 2006 and will have its North American premiere in September 2008 at the Soho Playhouse Theatre in New York City, states:

— *Terms such as "marriage," "family," and "parent" used to have firm denotations. They were the rock on which our cultural values rested. Terms such as "embryo," "baby," or "twin" were also considered unambiguous. Assumptions that marriage must be heterosexual and that a child cannot have two parents of the same sex were never even considered assumptions, because they were beyond questioning.*

— *All of these terms have become destabilized, their meanings blurred, their ranges extended. Some would blame in vitro fertilization technology during the past three decades for these developments, but in actual fact major social and cultural changes—primarily in the United States and Europe—were even more responsible for the monumental shift that has caused so much fear and antagonism, especially among the ever increasingly strident fundamentalists in the United States. So why not write a play about a situation where "family" and "parent" have assumed disturbingly fuzzy meanings? This is why I have situated* Taboos *in two of the socially and politically most polarized parts of the United States: the San Francisco Bay Area and the American Deep South.*[204]

[public blog]

CONCERNED HEART

— How about a discussion of the IVF technologies allowing older and older men as well as women to reproduce? What about the relationship between advancing paternal age and genetic/DNA mutations leading to de novo autism, de novo schizophrenia, etc., in the offspring?

CARL DJERASSI

— I have addressed this issue in my scientific publications, lectures, novels, and play (*An Immaculate Misconception*). I was positing the storage of young gametes—ovarian tissue or eggs by a woman in her twenties and sperm storage by a young man—to be used optionally in later years if childbearing is postponed to that time. Hence we are not talking about genetic material from aging parents. What needs to be debated, instead, are the pros and cons of the quality of parenting by older parents, in other words increased wisdom versus possibly impaired physical agility.

MIRIAM VAN RIJSINGE

— Dear Suzanne and all,

— As I came down with a very heavy flu, I was unable to contribute more, unfortunately. I will, however, have much to read and think about, as there are many sparkling points of interest that should be pursued further. Perhaps I will come back to some of the postings individually later, when I am well again.

— In the Netherlands we have the *Genesis* show at the Utrecht Central Museum, starting April 14th, with a symposium in

June. Bio-artist Adam Zaretsky[205] is doing his guest professorship at Leiden University in April and May, hosted by The Arts & Genomics Centre. Two more research projects (initiated by TA&GC) are financed and start this summer under the title "Imagining Genomics: Introducing Visuality in the Genomics Debate." It focuses on the word-image issue and on the function of the visual in the ethics debate. We will meet at various conferences, no doubt, and I am looking forward to it.

— I thank you, Suzanne, for this conference, and for your various leads (both words and images) in the discussion. It was quite an experience.

ANDREW CARNIE

— The symposium has to end just when I was beginning to get the hang of the whole event. Well done, JD Talasek, Suzanne, and all that were involved behind the scenes. Hope you get well soon, Jill. Many, many thanks but it is time to disperse *[fig. 66]*.

MICHAEL SAPPOL

— I want to congratulate the participants, JD Talasek, and the other organizers, and especially the moderator, Suzanne Anker, for an enjoyable and enlightening symposium.

— As for my projects, I'm currently working on two that may be of interest to the group:

— a DVD publication series of historical medical films from the collection of the National Library of Medicine *[fig. 67]*.
The first volume, on American World War II-era public health films, is very near completion; other volumes will be on cancer, tuberculosis, child development, human psychology, dental health, etc.

— *Man as Industrial Palace: Fritz Kahn, Conceptual Medical Illustration and the Visual Rhetoric of Modernity*, a book project on the origins of modernist medical illustration, focusing largely on the work of Fritz Kahn (1888-1968) *[fig. 68]*.

INGEBORG REICHLE

— Thanks to Suzanne for making this symposium possible, and thanks to JD Talasek and all the people who were involved behind the scenes.

— Next week we have a symposium, "Visual Models" (*"Visuelle Modelle"; "Fragen der Bildwissenschaft"*), in the context of image sciences here in Berlin, where some of the issues adressed by Suzanne and others will be on the agenda.

— I hope to see some of you at the workshop "Biomedicine and Aesthetics in a Museological Context" in late August 2007 in Copenhagen, followed by a one-day public conference on biomedicine and art.

— All the best!

SUZANNE ANKER

— Dear Panelists, Conference Hosts, IT Specialists, Bloggers, and JD Talasek,

— To all on board, my sincerest appreciation for your participation in our conference. Your thoughtful dialogue, opinionated comments, humor, and engaging ideas provided a stimulating and at times a well-needed, if not contentious, discussion. The online symposium proved itself to be a worthy form of exchange and allowed for a contagion of comments, not to mention the flu. The manner of our discussion incorporated personal style as a method of doing business, for better or for worse. Embedded in the results was the desire to let conversation do what it does as a hypertextual mode of connecting. As a self-generating system of yeas and nays, our conference ran its course of obstacles. However, I cannot report any casualties. As an update on current ideas at the nexus of visual culture and the biosciences, our dialogue will be added to the literature on this subject and hopefully be employed as a tool to generate further ideas, discourses, projects, and the like. I have never had the experience of being so ardently tuned into my computer. It reminded me of my teenage years, waiting in anticipation for the telephone to ring. And although our conversations were solely conducted through zeros and ones, I couldn't help myself feeling delightfully ever-present in the rush and pace of energized personas. Thank you all, and we will be in touch.

JD TALASEK

— On behalf of the NAS and UMBC, we wish to thank Suzanne Anker for her insightful moderation of this symposium and all of the panelists for their commitment to this discussion, both in the context of this conference and in their personal work. We would also like to thank the over 2,500 members of the Internet community who participated online during the conference.

1
Jock McDonald,
"The Father of the Pill," Dr. Carl Djerassi as a Pregnant Man, 1992, gelatin silver print, 7 x 7 inches.

2, 3, 4
Scenes from *An Immaculate Misconception,* performed at the Grutli Theater, Geneva, Switzerland, from April 12 to May 5, 2002. Written by Carl Djerassi, directed by Geoffrey Dyson, starring Caroline Gasser, Michel Kullmann, and Roberto Salomon. Photography by Valdemar Verissimo. www.photoverissimo.com.

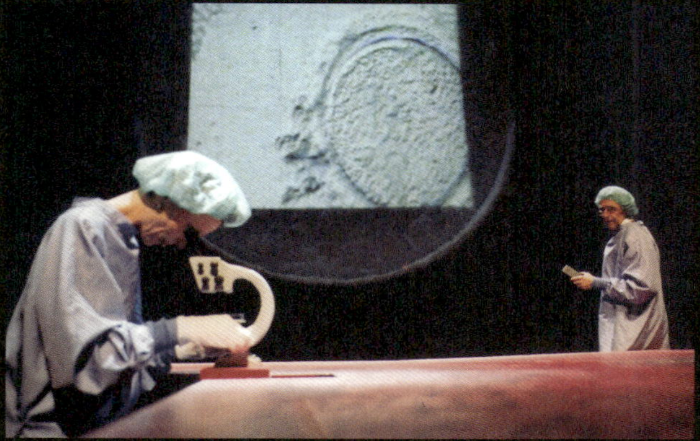

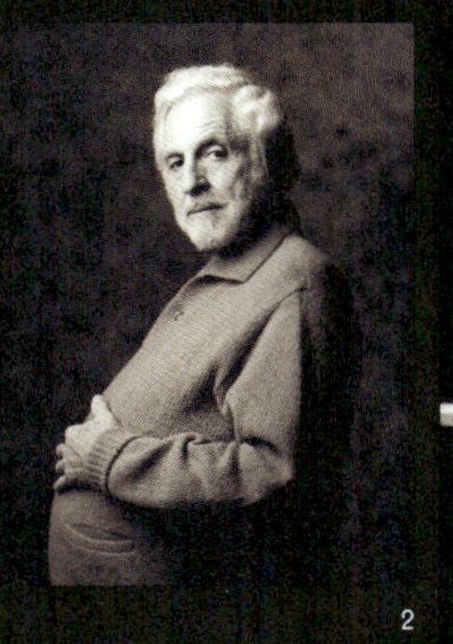

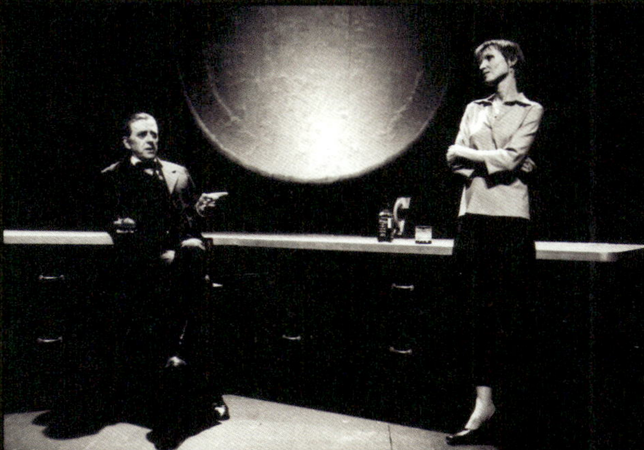

2
1 3
4

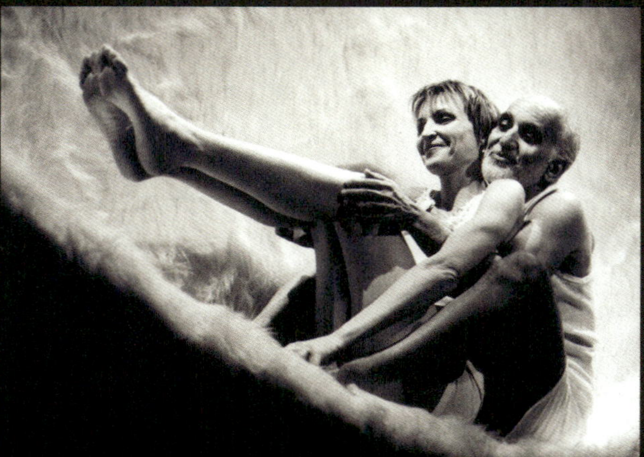

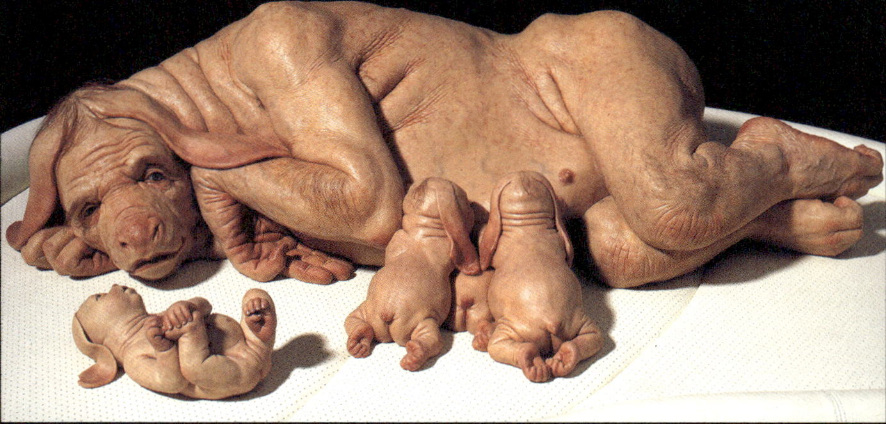

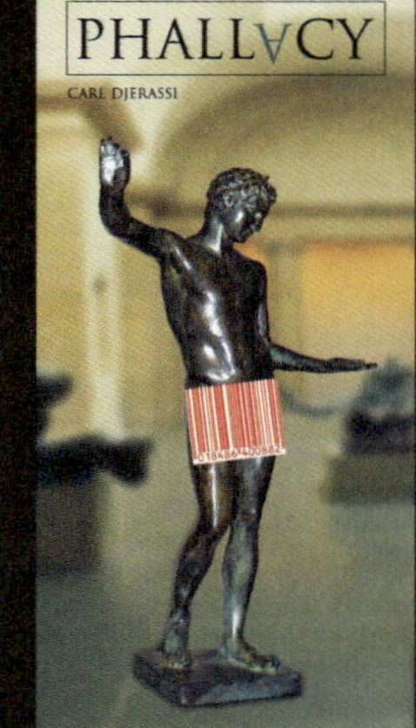
PHALLACY
CARL DJERASSI

7
An image from the audio-visual component of Carl Djerassi's play *Phallacy.*

8
Ludwig Krug, *Man of Sorrows*, c.1520, engraving, 11.8 x 7.8 inches.

9
Nicholas Nixon, *Suzanne Richardson, Boston*, 2005, gelatin silver print, 8 x 10 inches. © Nicholas Nixon, courtesy Fraenkel Gallery, San Francisco.

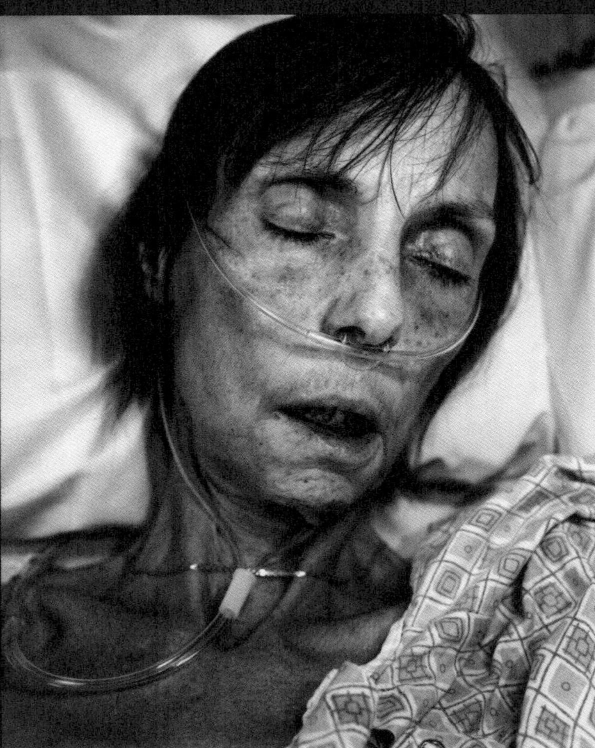

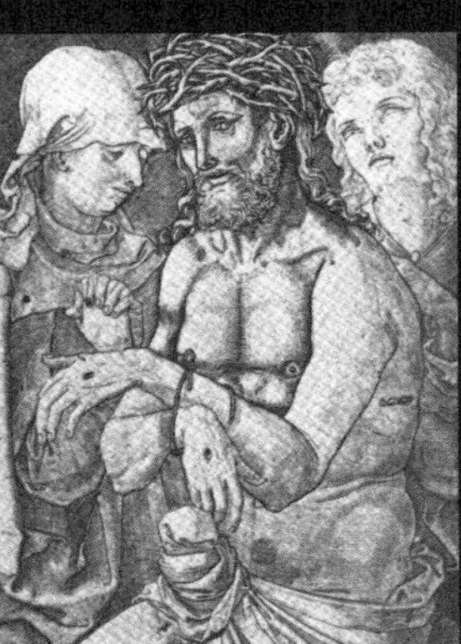

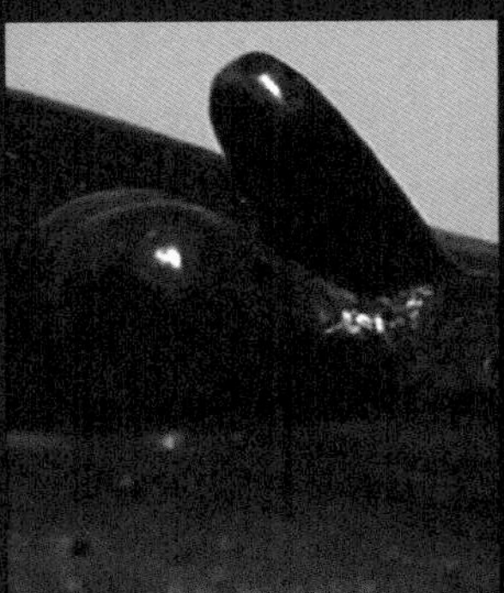

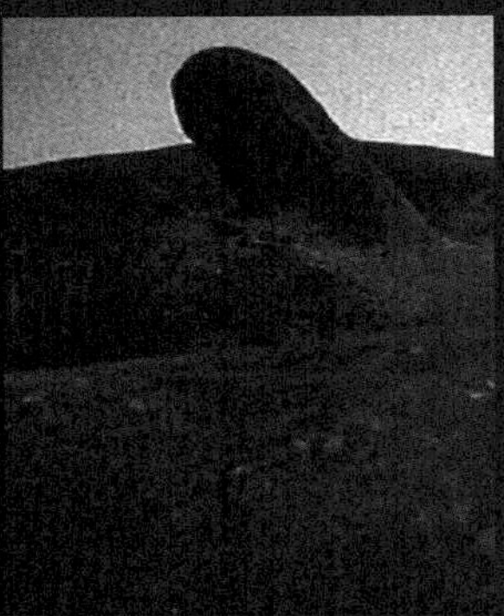

9
7 8

10
Giovanni Frazzetto, *Qualia Bar*, 2004 to 2006, light box with colored liquid. Courtesy of Giovanni Frazzetto. Photography by Henry Sanchez.

11
Giovanni Frazzetto, *Qualia Bar* (detail), 2004 to 2006, light box with colored liquid. Courtesy of Giovanni Frazzetto. Photography by Henry Sanchez.

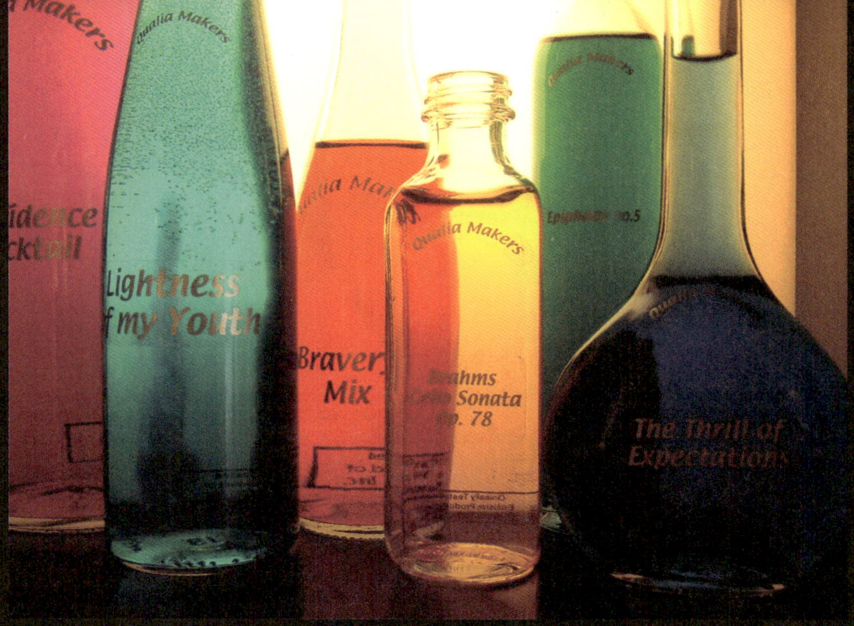

11

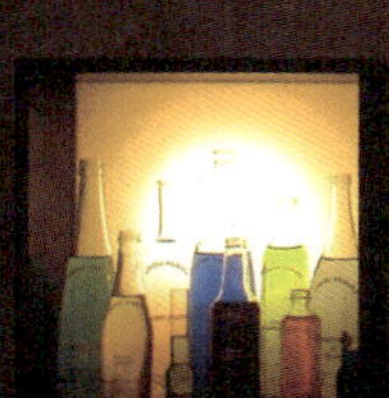
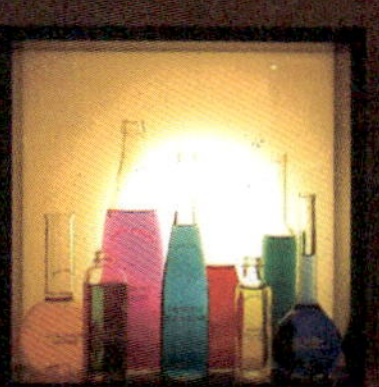
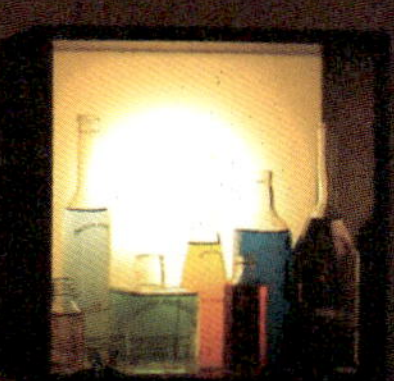

10

12
Jill Scott, *The Mediated Stage* from the project *e-skin-designing tactile interfaces for visually impaired on the mediated stage*, 2006, Human Computer Interface, Cooperation between the Zurich University of the Arts and the Artificial Intelligence Lab, University of Zurich.

13
Jill Scott, *The Electric Retina*, 2008, interactive media, 78 x 47 inches. Sculptural commission about the neuro-morphology of the visual system for the Brain Fair. University of Zurich.

14
The photoreceptor pattern array of Zebrafish Larva, 2007, tangential protein staining. Courtesy of the Institute of Zoology. Neurobiology, University of Zurich.

15
Histology of Wild Type Zebrafish Retina, 2006, microscopic photography. Courtesy of the Institute of Zoology, Neurobiology, University of Zurich.

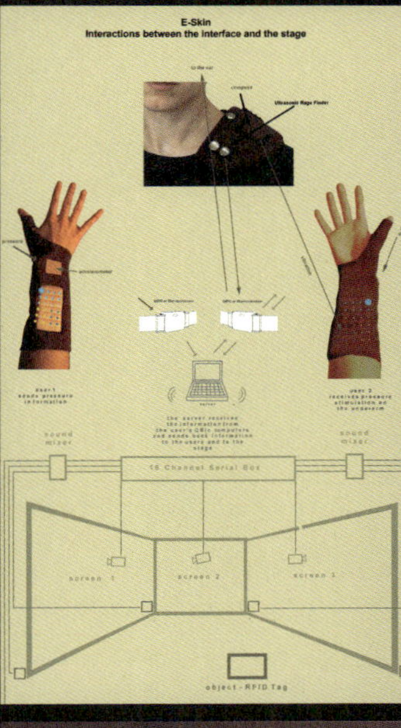

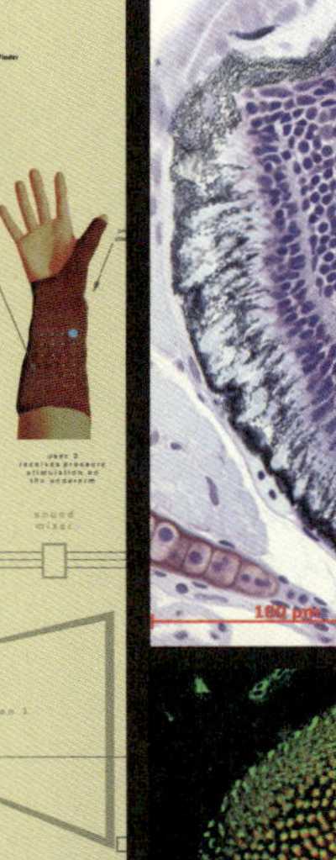

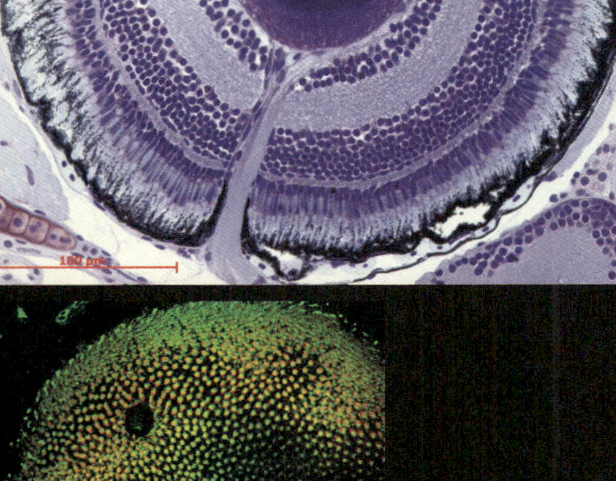

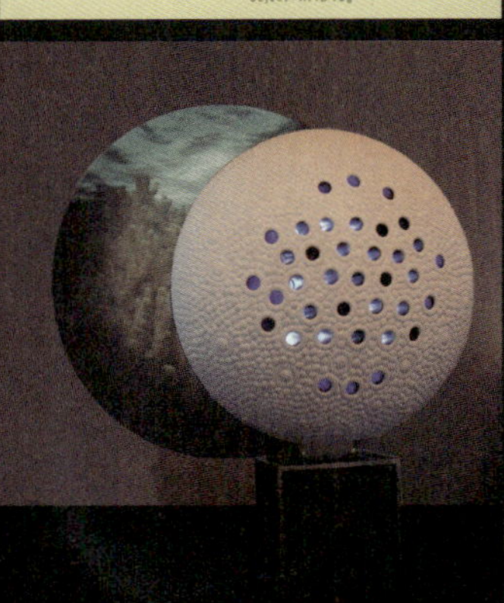

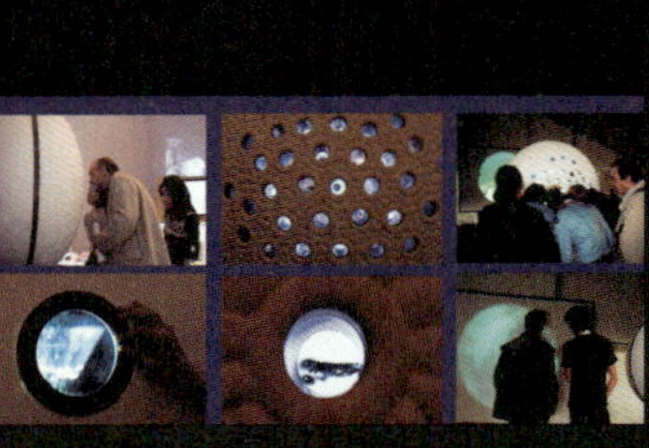

12 15
13 14

Jean Pagliuso,
Black #1, 2005,
gelatin silver print on Thai Mulberry paper, 19 x 23.75 inches. Courtesy of Jean Pagliuso Photography.

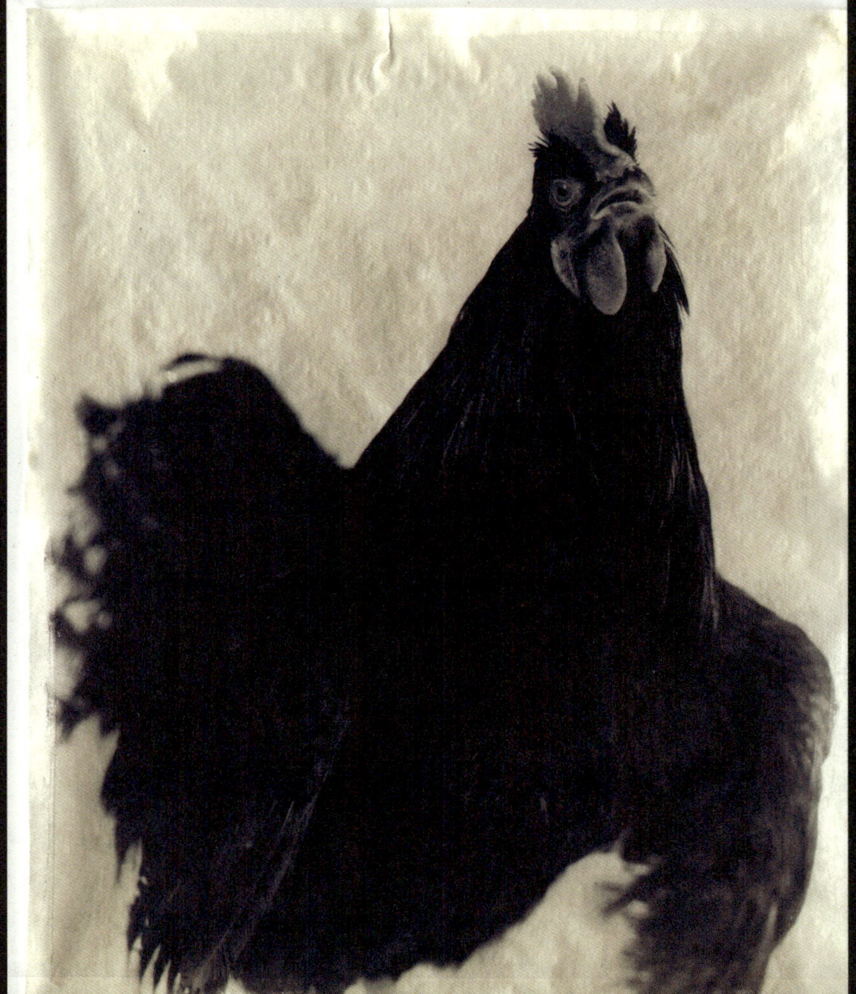

17
Max Aguilera-Hellweg, from *The Sacred Heart*, removal of plaque from the carotid artery, 1997.

18
B. Kaimovitz, Y. Lanir, T. Wischgoll, and G.S. Kassab, *Swine Heart*, 2003-2007, CAD blueprint. Courtesy of the Department of Biomedical Engineering at Indiana University–Purdue University, Indianapolis.

19
Max Aguilera-Hellweg, *Infant with Midline Facial Dysplasia*, Bolivia, 2000, 4x5 inch Fujifilm RDPII.

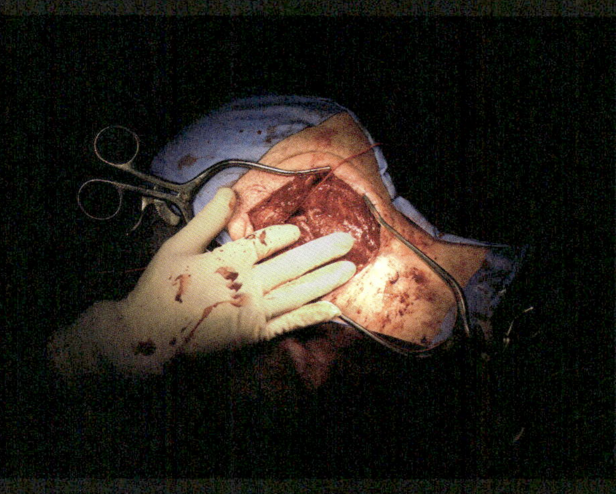

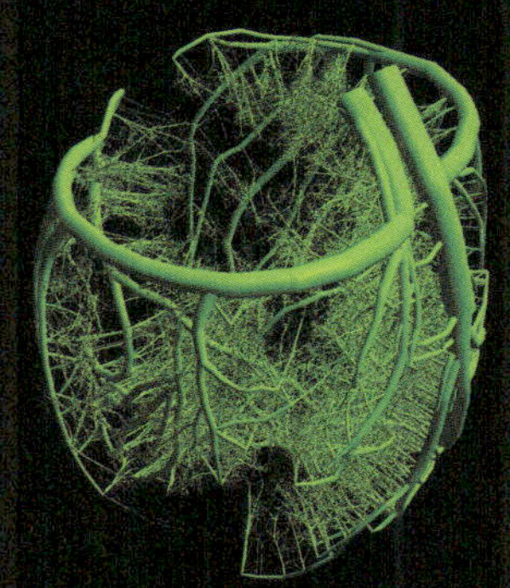

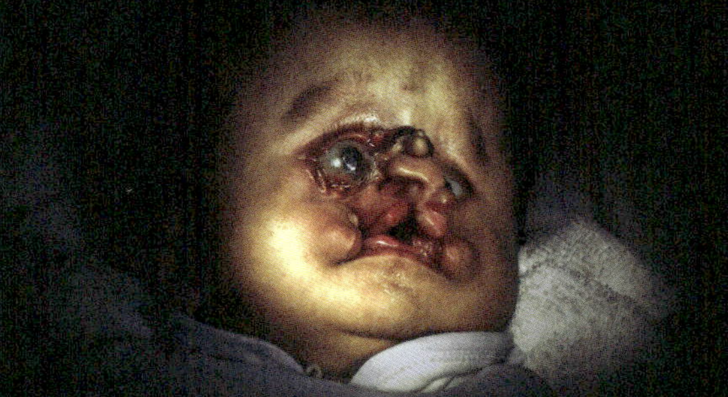

people find me disturbing. in the hospital, they even put gauze
over my eyes and face so no one would have to see me. when i
feed on my mother's tit, i can't get enough to eat because
my mouth just isn't right. i'm just a mess. but if you closed
your eyes and picked me up in your arms you'd find i'm warm
and cuddly. i'm just a boy, just a little boy, don't be afraid.
you can look, i won't bite, i can't, i don't have a mouth...

17 18
19

20, 21, 22
Andrew Carnie, *Magic Forest*, 2002, slide dissolve installation, 120 x 120 x 240 inches, shown at the Science Museum, London, United Kingdom. Photography by Andrew Carnie.

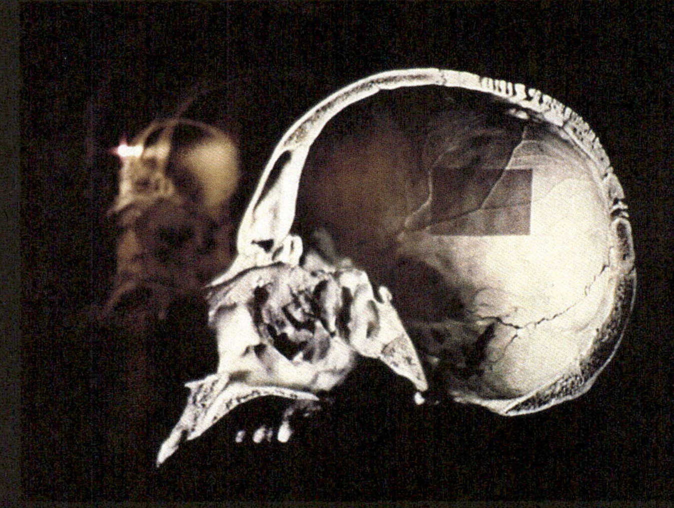

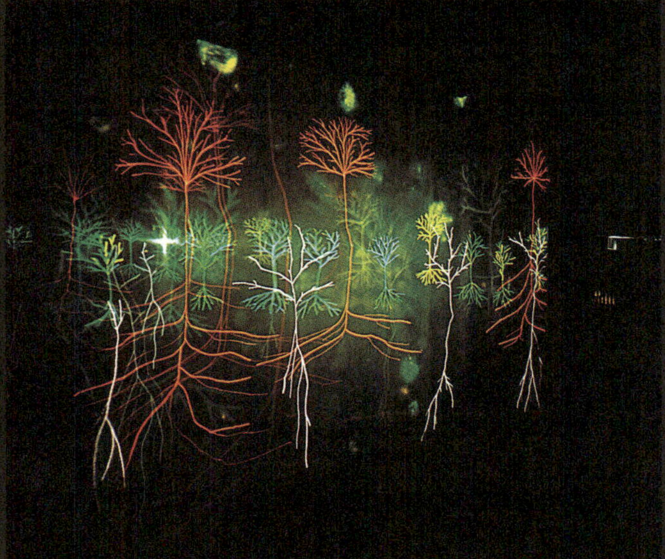

23
Julian Gabriel Richter, Christopher Roman in the Forsythe Company's production of *You Made Me a Monster* at the Festival d'Avignon, France, 2005, courtesy of Julian Gabriel Richter and The Forsythe Company.

24
Marion Rossi, Nicole Peisl in the Forsythe Company's world premiere of *You Made Me a Monster* at the Venice Biennale, 2005, courtesy of Marion Rossi and The Forsythe Company.

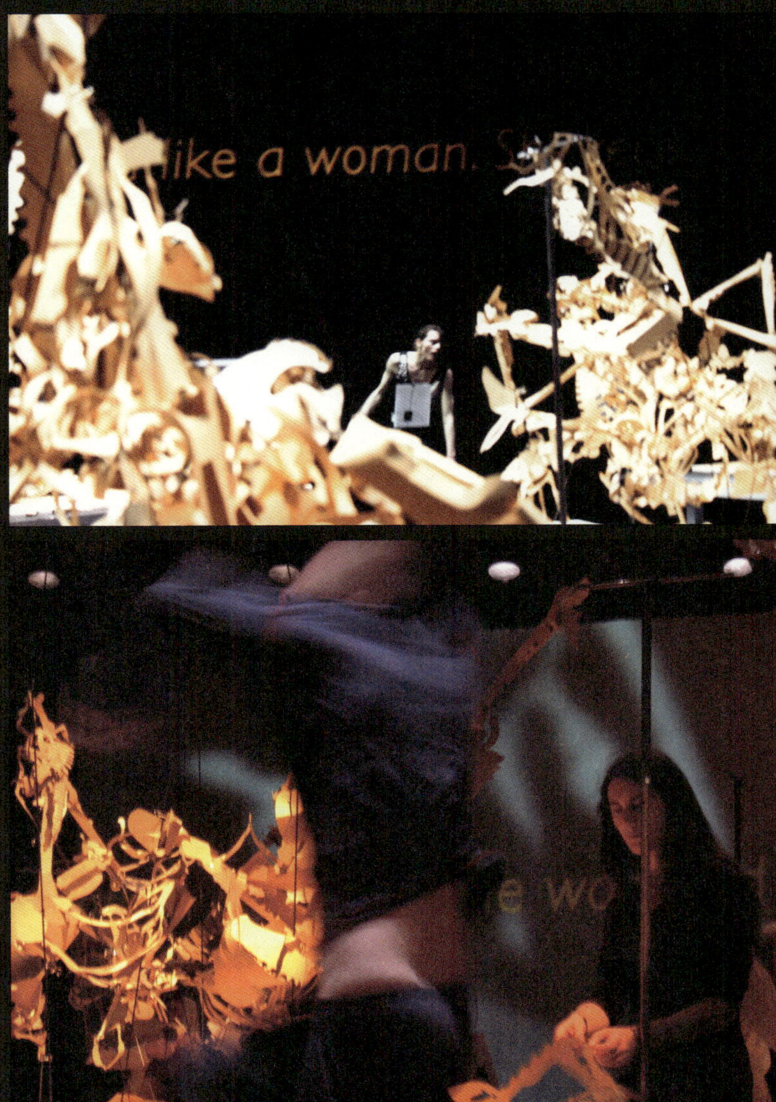

Andrea Vesalius, *De humani corporis fabrica libri septem*, Basel, Johannes Oporinus, 1543, woodcut, Book V, fig. 22. Collection of the National Library of Medicine.

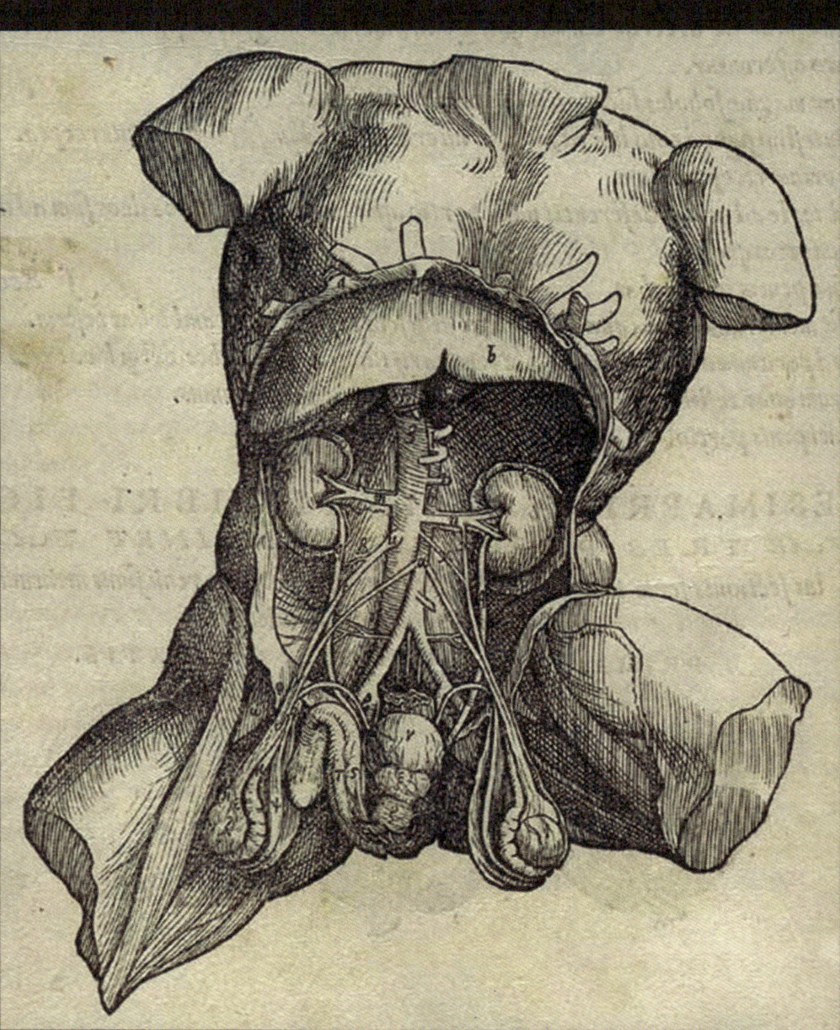

26
Austrian stamp issued in 2005 to honor Carl Djerassi. Courtesy of the Austrian Postal Service.

CHEMIKER / ROMANCIER
1923 GEBOREN
1938 VERTRIEBEN
2003 VERSÖHNT

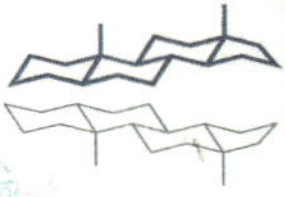

CARL DJERASSI

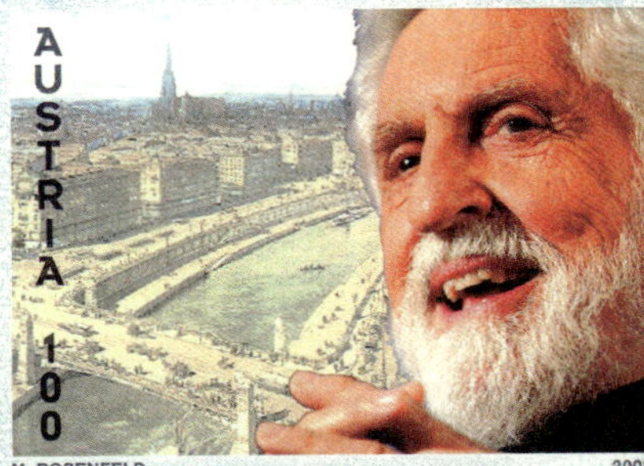

M. ROSENFELD 2005

27, 28
Brian Fies, Pages 54 and 53 from *Mom's Cancer*, 2006, India ink on Bristol board, 9 x 12 inches.

CANCER IS THREE-DIMENSIONAL. IT HAS **VOLUME.** SO WHEN A TUMOR'S LENGTH SHRINKS FROM, SAY, 5 INCHES TO 4 INCHES, THAT'S REALLY A VOLUME DIFFERENCE OF $4^3 \div 5^3 = 64 \div 125 =$ ABOUT ONE-HALF.

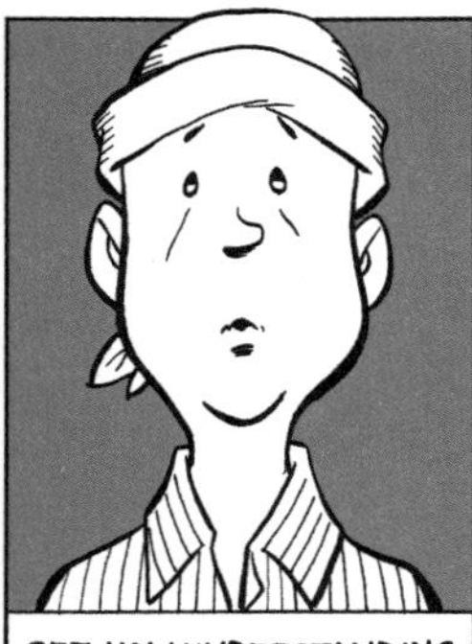

SEE HOW UNDERSTANDING MATH HELPS? SEE HOW WHEN THE DOCTOR SAYS:

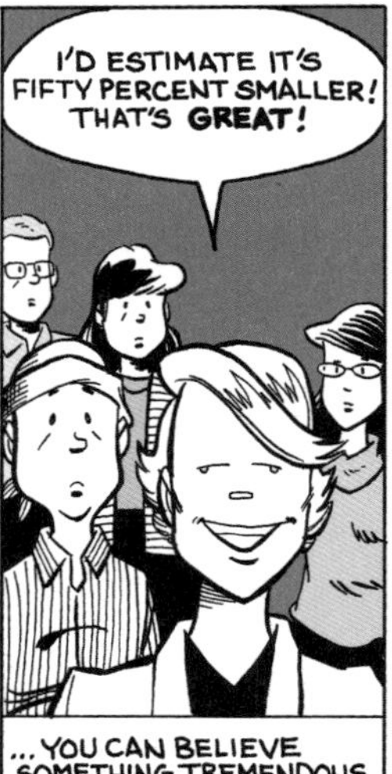

...YOU CAN BELIEVE SOMETHING TREMENDOUS HAS BEEN WON? SEE HOW YOU CAN LEAVE THE OFFICE HAPPY AND PROUD?

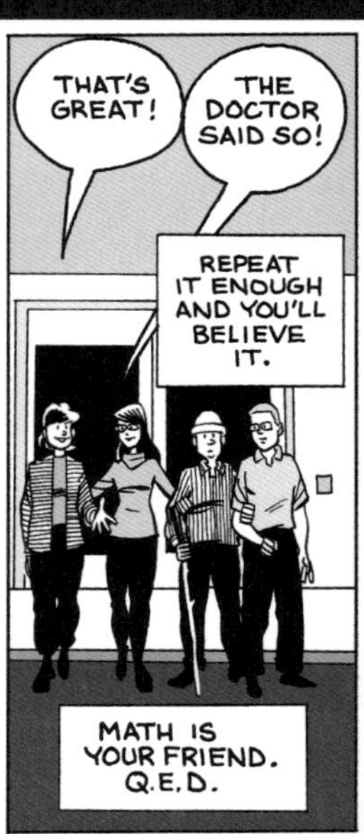

MATH IS YOUR FRIEND. Q.E.D.

ONE DIMENSION:
LINE A IS **2 TIMES** AS LONG AS LINE B.

A B

TWO DIMENSIONS:
AREA VARIES WITH THE SQUARE OF LENGTH. THE AREA OF SQUARE A IS 2^2 = **4 TIMES** THAT OF SQUARE B.

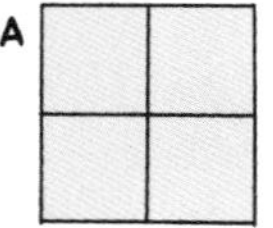

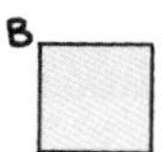

THREE DIMENSIONS:
VOLUME VARIES WITH THE CUBE OF LENGTH. THE VOLUME OF CUBE A IS 2^3 = **8 TIMES** THAT OF CUBE B.

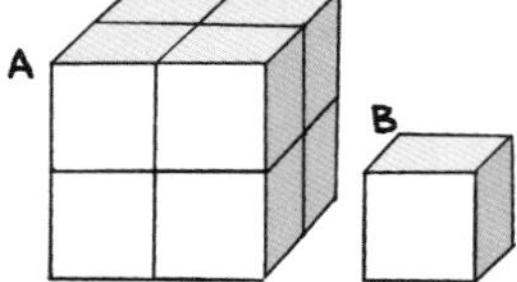

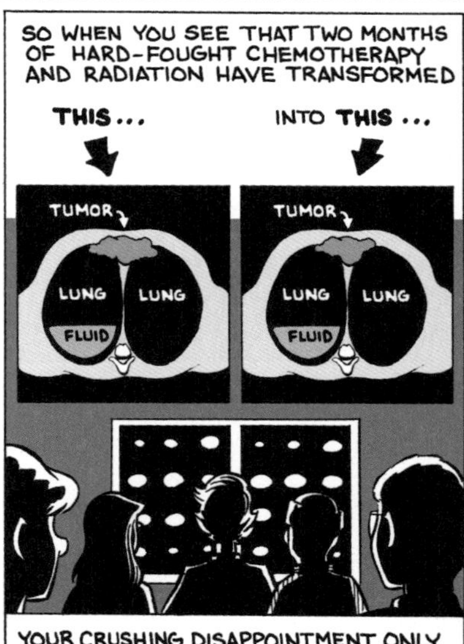

29, 30, 31
Suzanne Anker,
Rorschach (Wolf), Rorschach (Crab), and Rorschach (Bear), 2004, rapid prototype sculpture, plaster and resin, 5.5 x 5.5 x 2 inches.

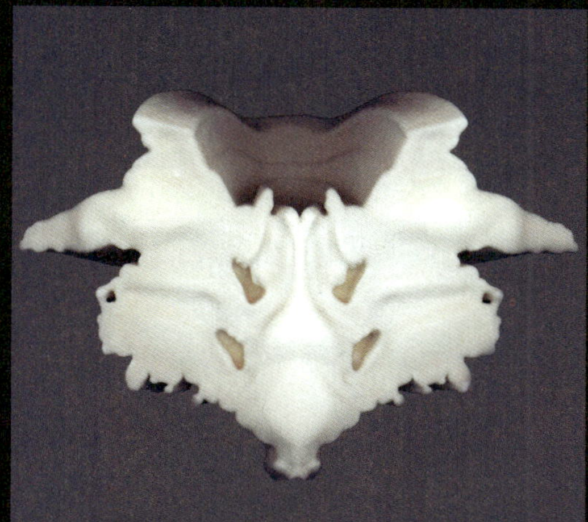

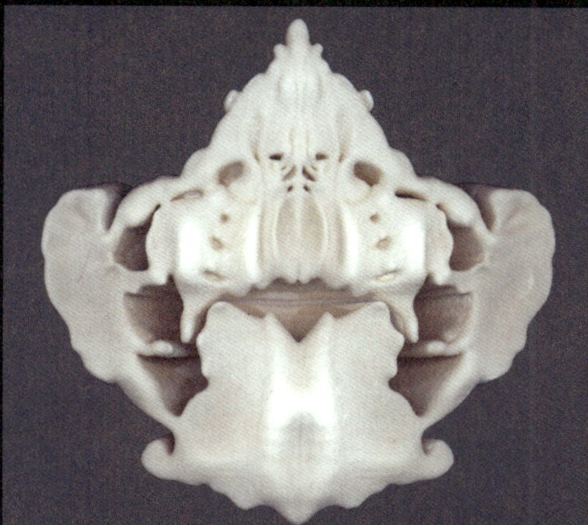

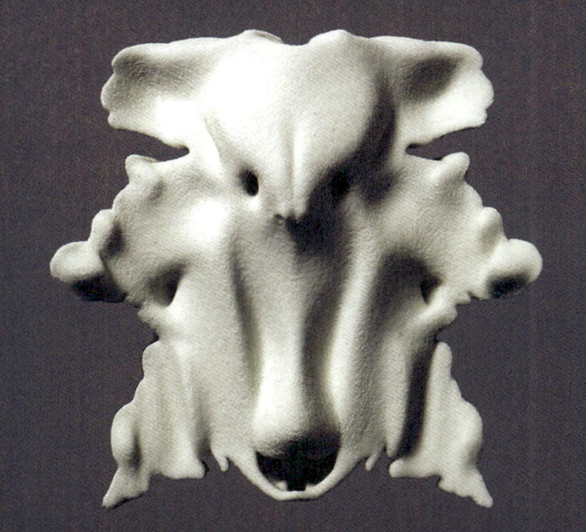

32
Leonel Moura, *ISU170906*, painting by the robot ISU (with its signature), 2006, acrylic on canvas, 30 x 30 inches. Courtesy of LEONEL MOURA ARTe Gallery.

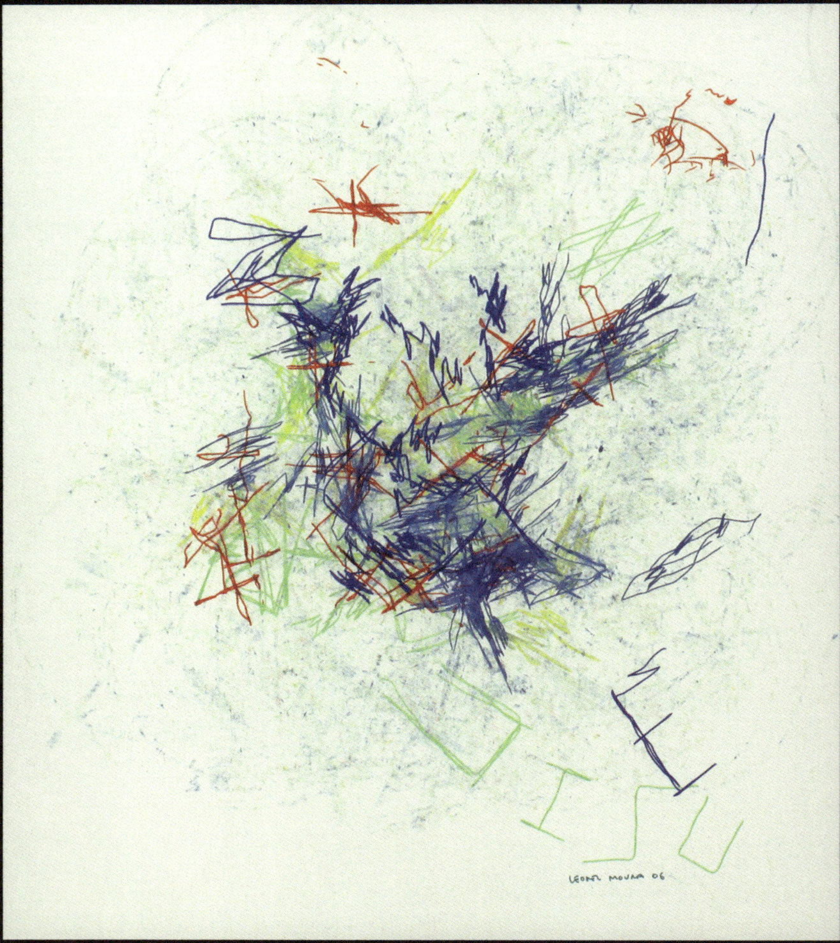

33
Brad Smith, *32-Day Human Embryo*, 2006, photomicrograph. Courtesy of Brad Smith, University of Michigan.

34
Time (November 11, 2002), image: Alexander Tsiaras, Time & Life Pictures/ Getty Images.

35
Newsweek (June 9, 2003), image: Steve Allen, The Image Bank / Getty Images.

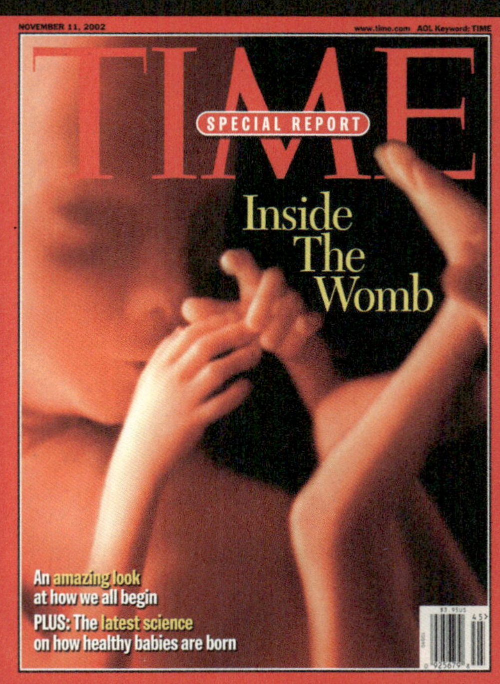

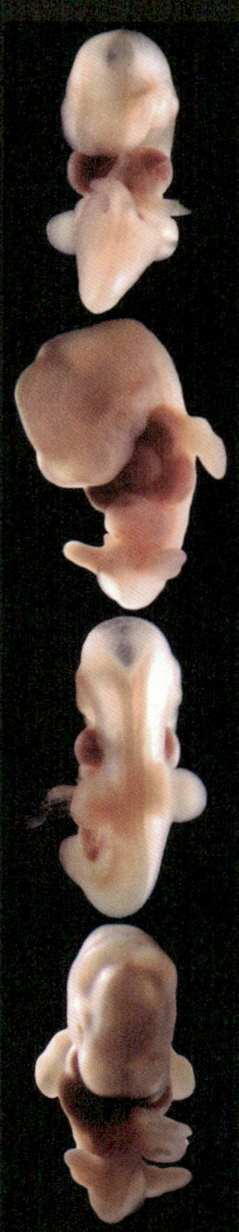

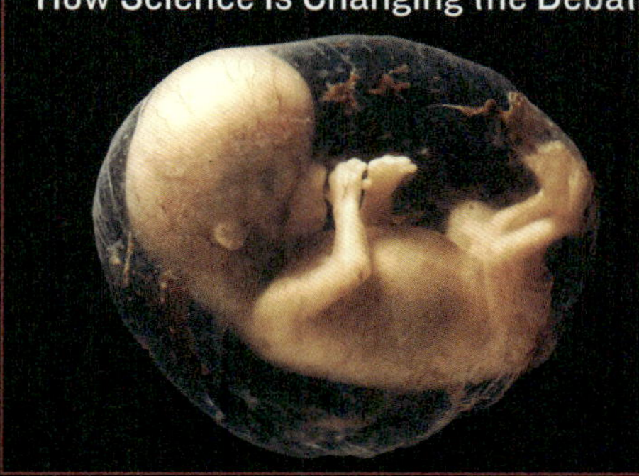

34 33
35

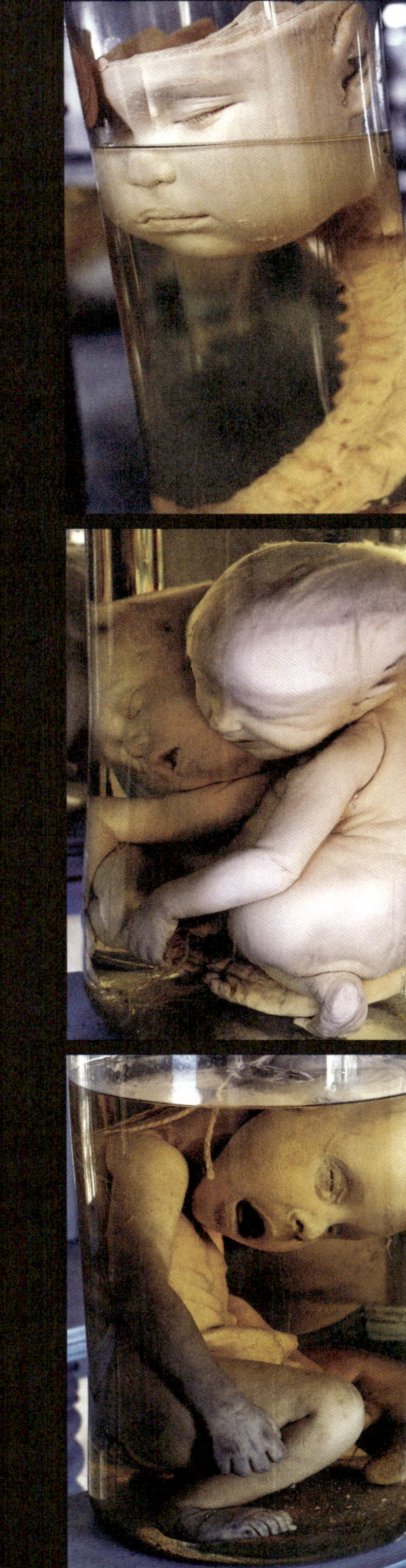

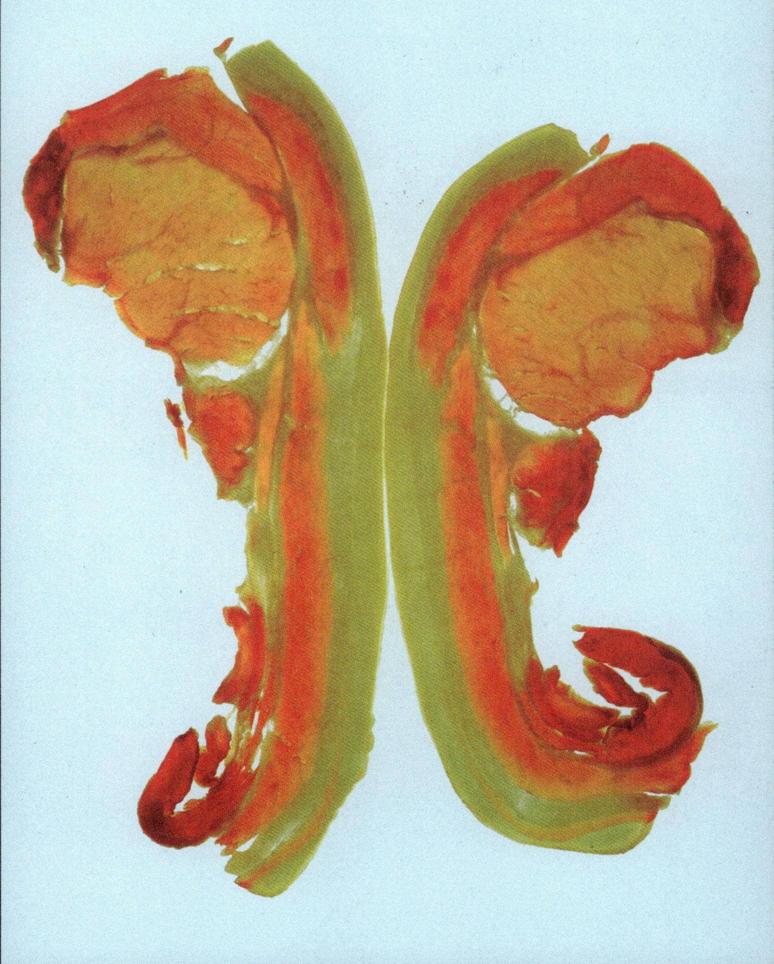

40
ORLAN, *Refiguration / Self-Hybridization*, African series: Fang Initiation Mask, Gabon, and Photo of Euro-Stephanoise Woman, 2002, Lambda print, 49.25 x 61.5 inches. Technical assistance: Jean-Michel Cambilhou.

41
ORLAN, *Refiguration / Self-Hybridization,*, American Indian series #1: painting portrait of No-No-Mun-Ya, One Who Gives No Attention, with ORLAN's photographic portrait, 2005, digital print, 60 x 49 inches.

42
ORLAN, *Refiguration / Self-Hybridization*, Precolumbian series, n°5, 1998, Cibachrome, 59 x 39 inches. Technical assistance: Pierre Zovilé.

43
ORLAN, *ORLAN with Her Multicultural Fairy Hat*, 5th Surgery-Performance titled *Operation-Opera*, Paris, 1991, Cibachrome, 69 x 47.5 inches.

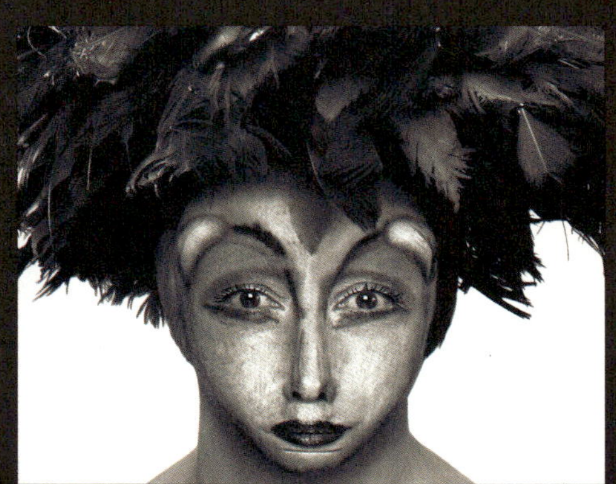

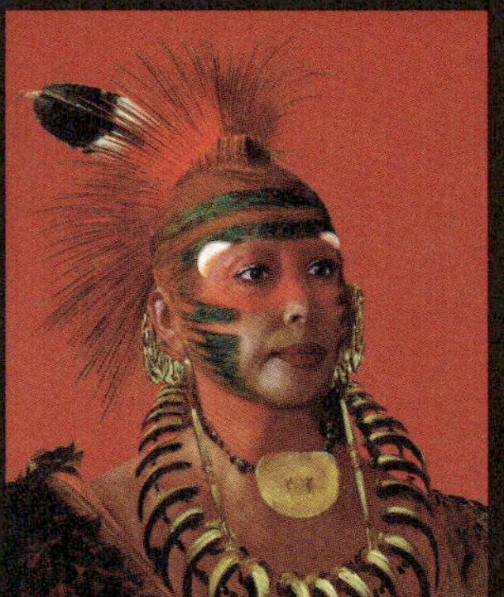

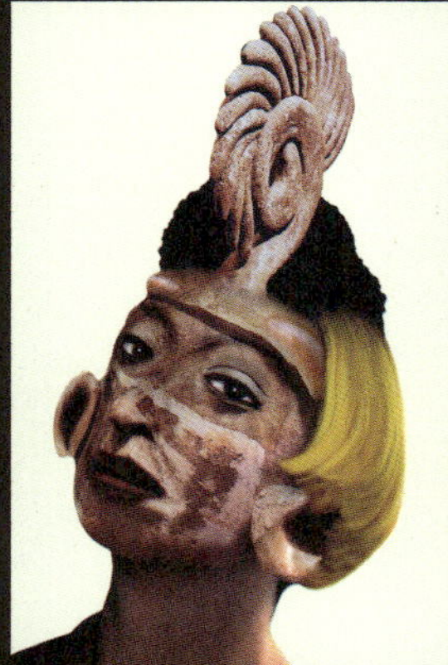

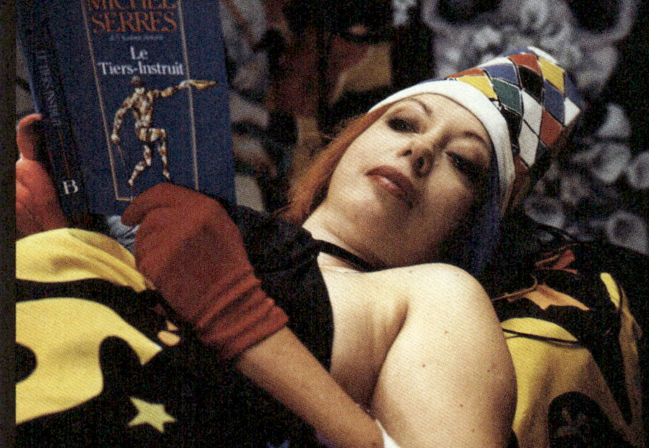

44
ORLAN, *Reliquary, My Flesh, The Text and the Languages*, steel frame, bullet proof glass, 10 grams of ORLAN's flesh in a cast of resin, 39 x 39 x 4.7 inches.

45
The operating-room during the biopsy, University of Western Australia, Perth, 2007.

46
ORLAN in harlequin costume on the biopsy table, University of Western Australia, Perth, 2007. Costume: Lucas Bowers, ericaamerica.

47
ORLAN, *Harlequin Coat*, 2007, video projection and bio-reactor with ORLAN's skin cells, black woman's cells and marsupial cells at the *Still, Living* exhibition, Bakery Artrage, Biennale of Electronic Arts Perth (BEAP).

..."WHAT CAN THE COMMON MONSTER, TATTOOED, AMBIDEXTROUS, HERMAP HRODITE AND CROSS-BRED, SHOW TO US RIGHT NOW UNDER HIS SKIN? YES, BLOOD AND FLESH. SCIENCE TALKS OF ORGANS, FUNCTIONS, CEL LS AND MOLECULES TO ACKNOWLE DGE THAT IT IS HIGH TIME THAT ON E STOPPED TALKING OF LIFE IN THE LABORATORIES BUT SCIENCE N EVER UTTERS THE WORD FLESH WH ICH, QUITE PRECISELY, POINTS OU T THE MIXTURES IN GIVEN PLACE OF THE BODY, HERE AND NOW, OF MUS CLES AND BLOOD, OF SKIN AND HAI R, OF BONES, NERVES AND OF THE V ARIOUS FUNCTIONS AND WHICH HEN CE MIXES UP THAT WHICH IS ANALIZE

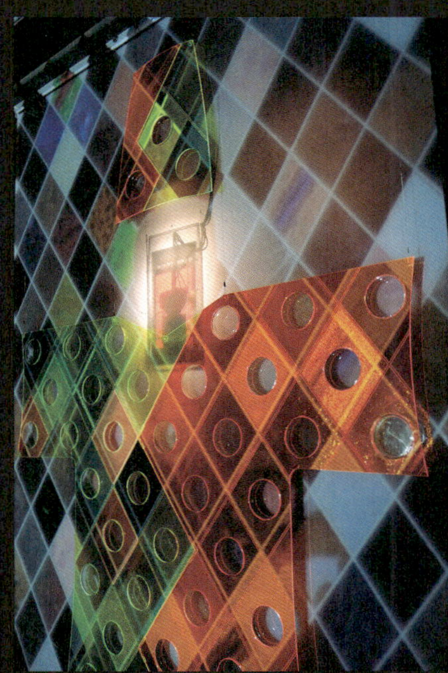

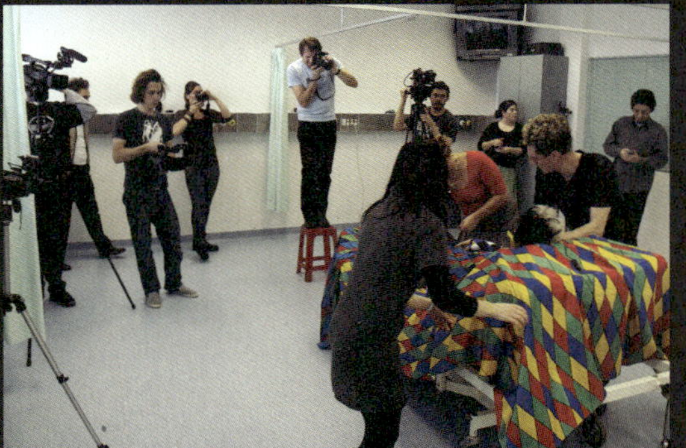

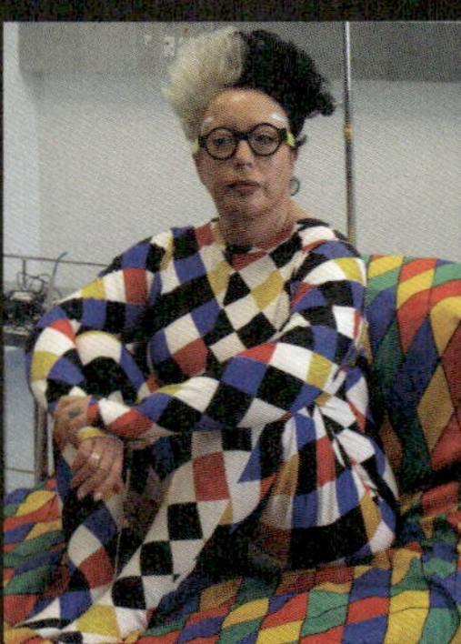

44 47
45 46

48
Fritz Lang, *Rotwang in his laboratory from Metropolis*, 1926. Rights: Friedrich-Wilhelm-Murnau-Stiftung; Distributor: Transit Film GmbH.

49
Chimera Plush.
Courtesy of Toy Vault, Inc.

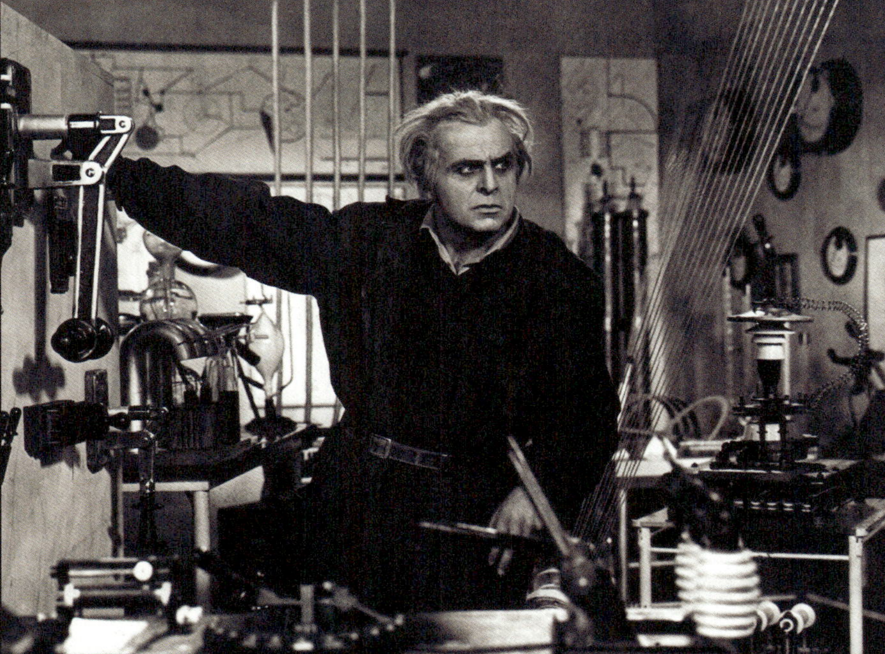

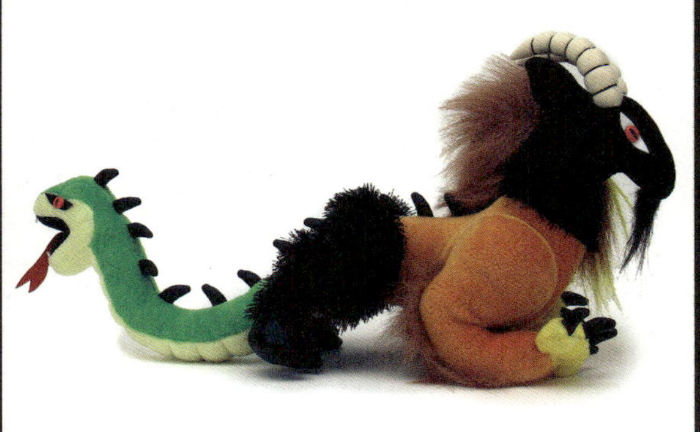

50
Oron Catts and Ionat Zurr, *Victimless Leather: A Prototype of a Stitch-less Jacket Grown in a Technoscientific "Body,"* 2004, biodegradable polymer connective and bone cells, dimensions vary depending on the installation. Courtesy of the Tissue Culture & Art Project.

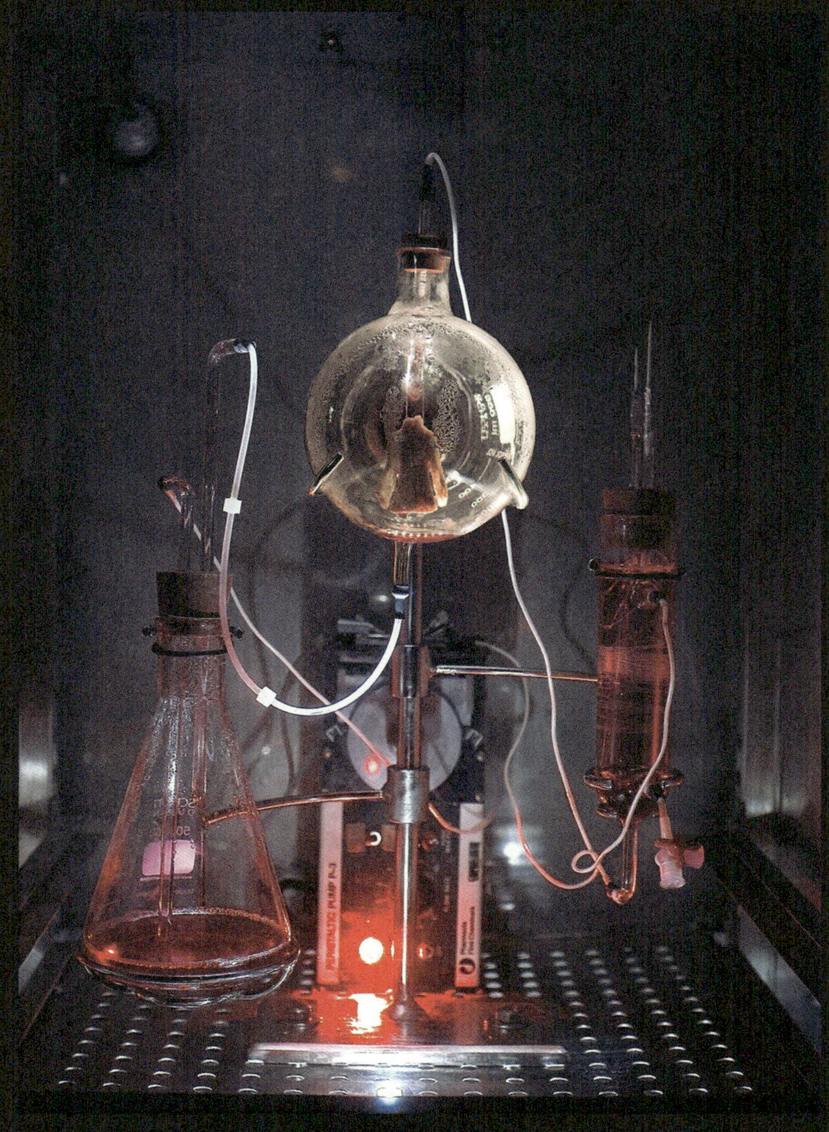

51, 52, 53, 54
Andrew Carnie, *We Are Where We Are*, 2006, eight-projector slide dissolve work, 120 x 480 x 480 inches. Made for the Art and Mind Festival, Winchester, United Kingdom. Photography by Andrew Carnie.

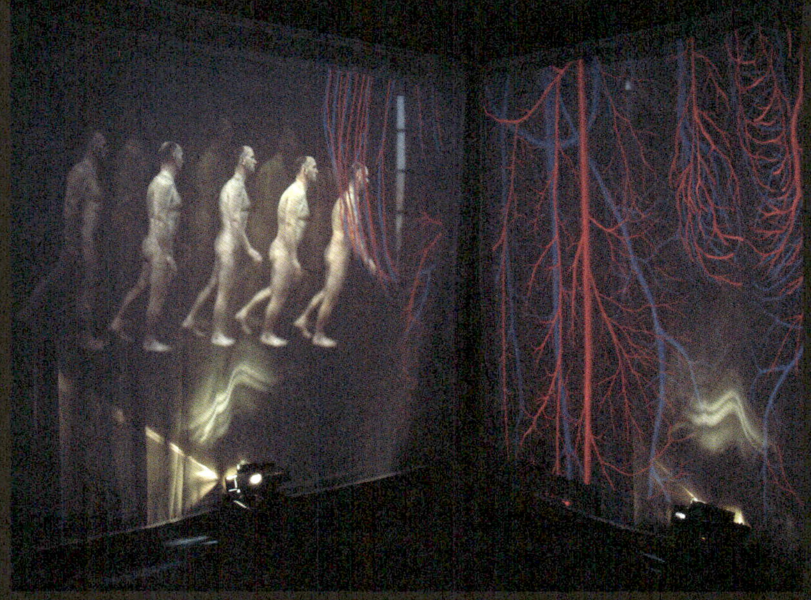

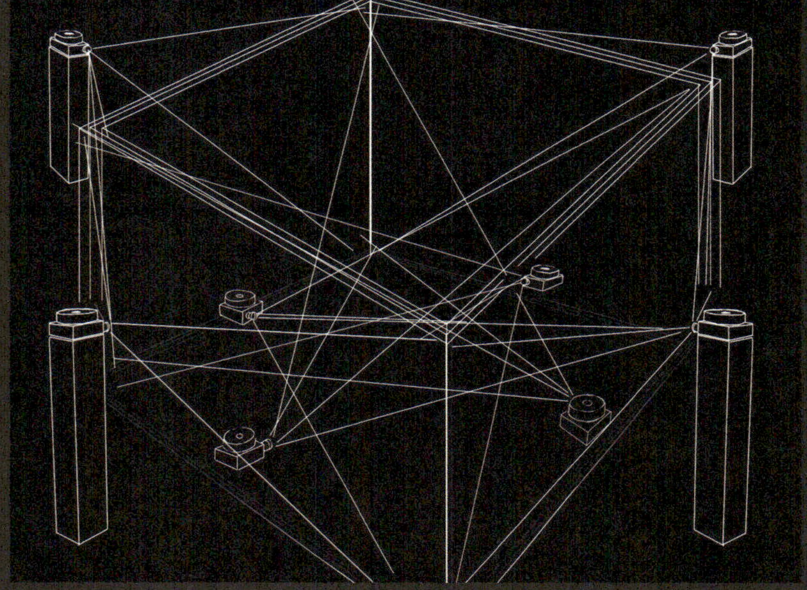

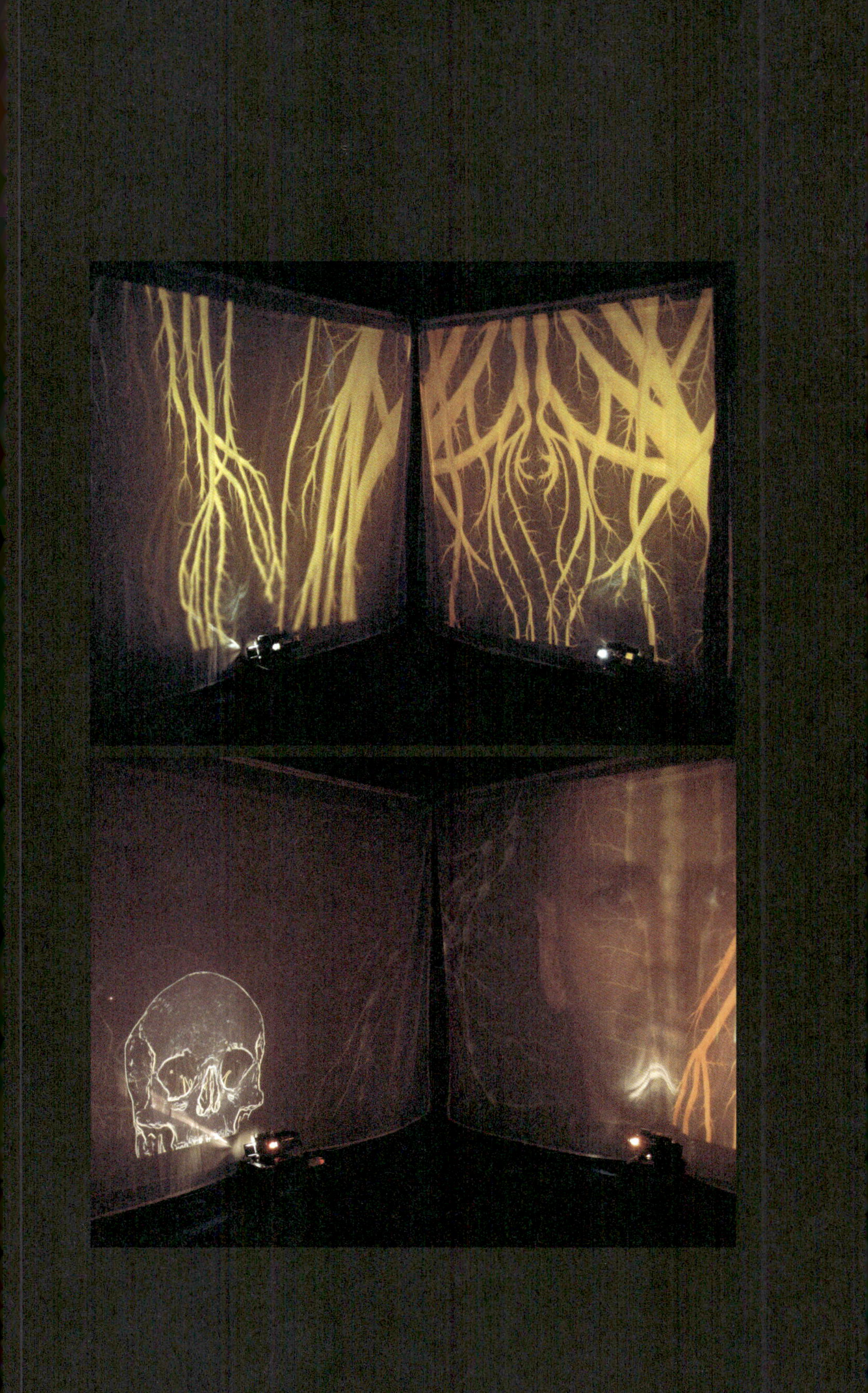

55
Eugenical Sterilization Legislation in the United States, 1921. Courtesy of the Cold Spring Harbor Laboratory Archives.

56
Catherine Chalmers, *Drinking*, 2000, chromogenic print, 60 x 40 inches. Courtesy of the artist.

57
Thomas Grünfeld, *Rosella/Mole from the series Misfits*, 2007, taxidermy, 4 x 6 x 2 inches. Courtesy of Thomas Grünfeld. © 2008 Artists Rights Society (ARS), New York / VG – Bild-Kunst, Bonn.

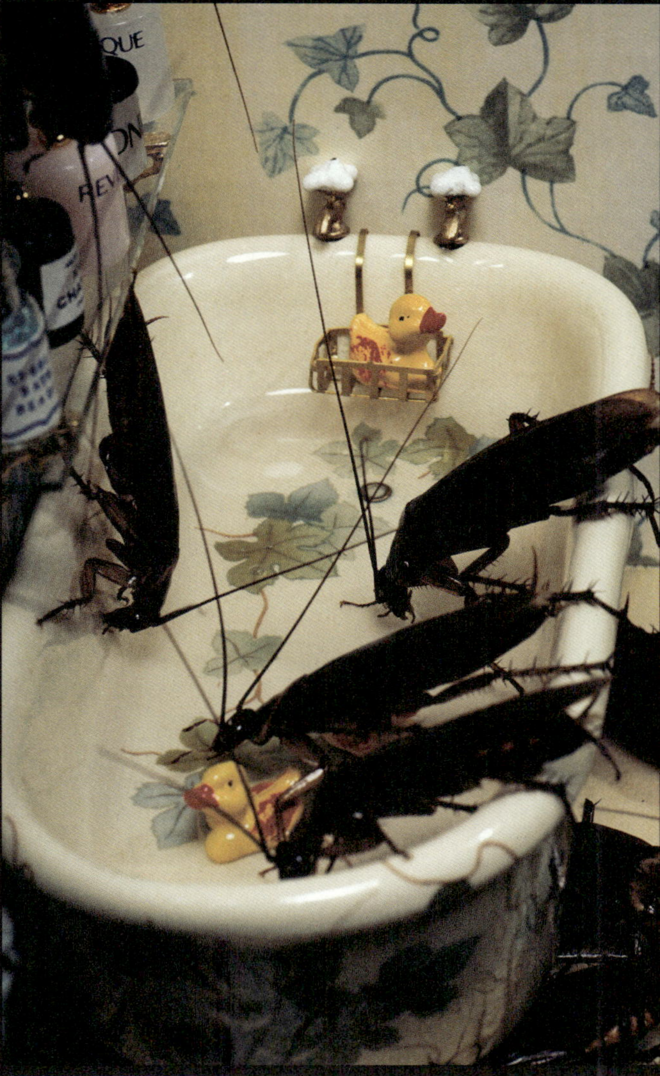

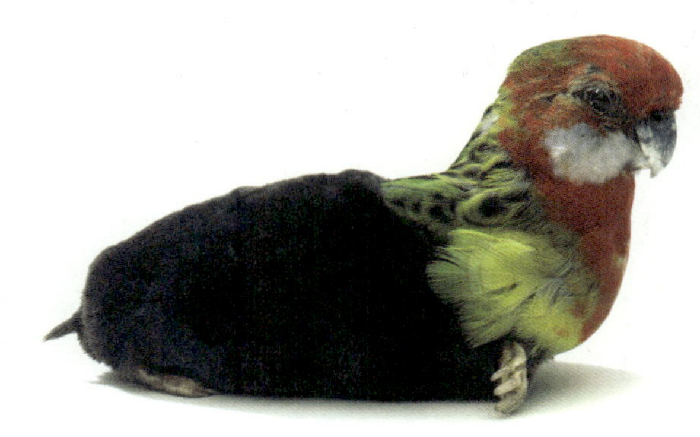

58
The London Zoo chimpanzee Congo achieved celebrity status as a painter in the 1950s, August 1957. Photo: John Pratt, Hulton Archive, Getty Images.

59
Congo the chimpanzee, *Untitled Abstract,* 1957, tempera on paper, 13.75 x 17.75 inches, courtesy of Desmond Morris. From the collection of Howard Hong, Arcadia, California.

60
Leonel Moura, *080404,* from the ArtSBot (Art Swarm Robots) series, 2004, acrylic on plexiglas, 39 x 39 inches. Courtesy of LEONEL MOURA ARTe Gallery.

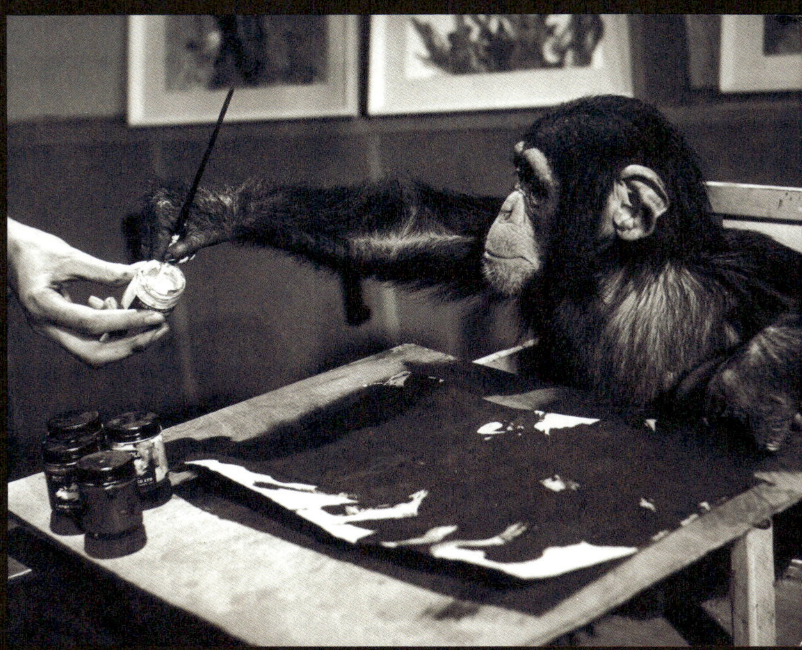

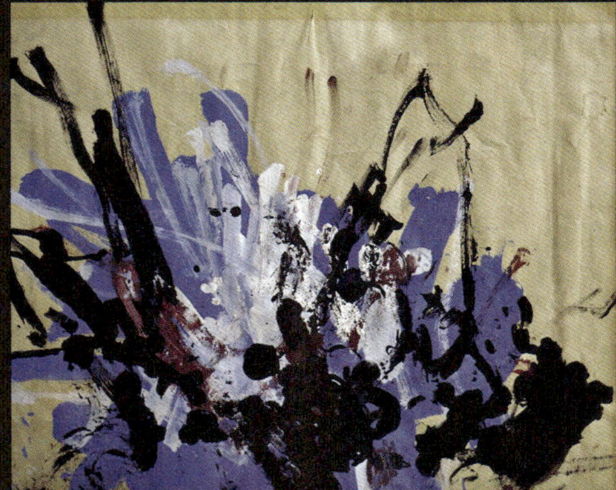

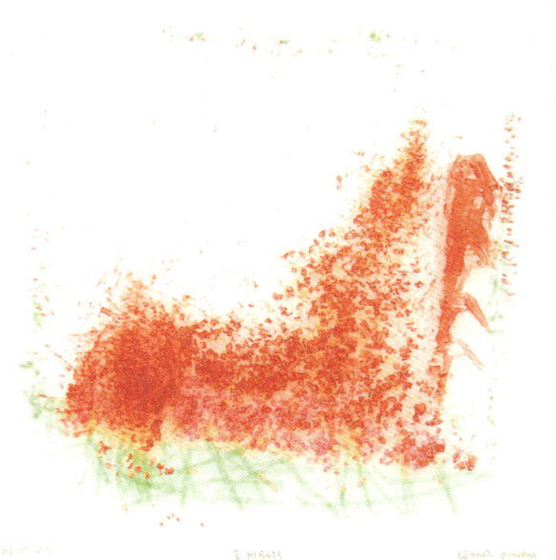

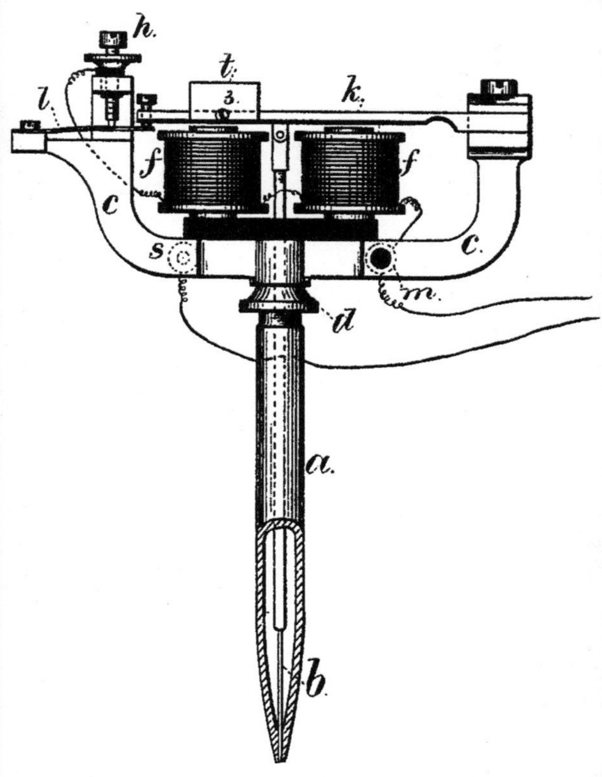
h.
t.
3.
k.
l.
f
f
c
c.
s
m.
d
a.
b.

62, 63
Andrew Carnie,
Things Happen, 2002,
slide dissolve work,
80 x 80 x 240 inches.
Made for the Mendel
Museum, Brno Czech
Republic. Photography
by Andrew Carnie.

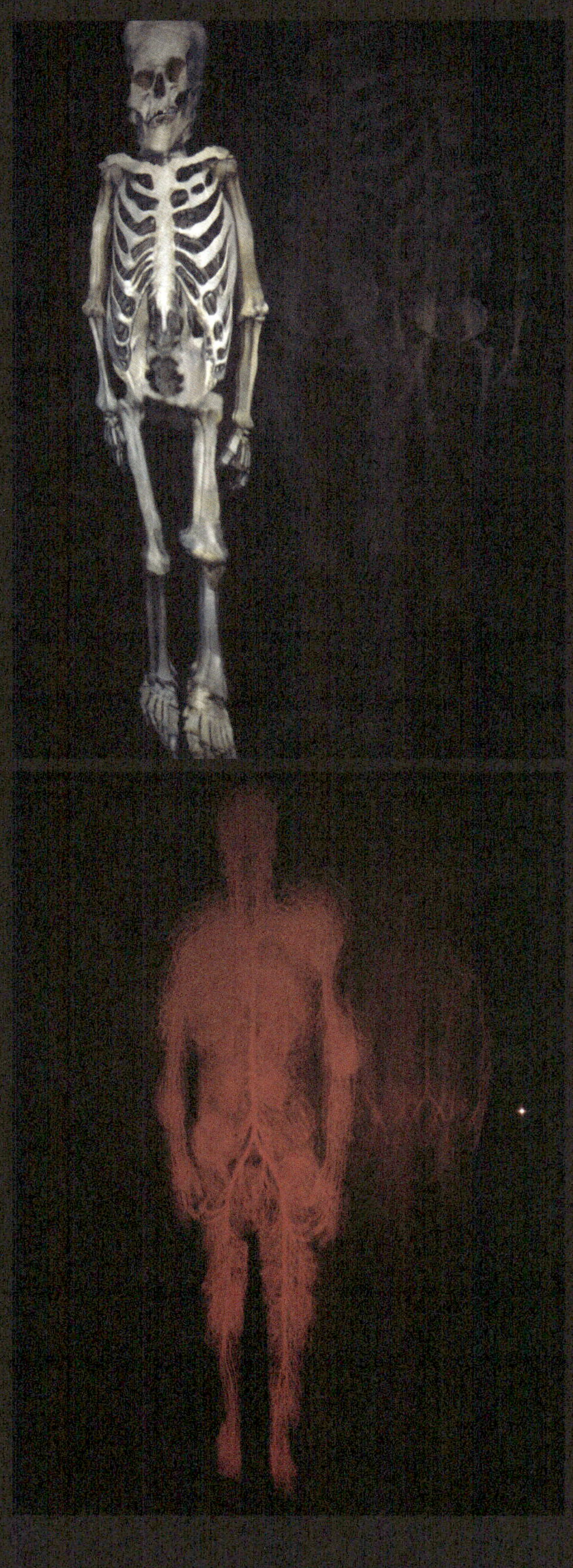

NINE PERCENT EIGHT HUNDRED AND EIGHTY-ONE THOUSAND SIX HUNDRED AND SEVENTY NINETEEN THREE HUNDRED AND FORTY-ONE ?

animals en Redheads - web-log.nl ... heart broken by a redhead because they may just call redheads evil bitches, but they are just bitter. ... The MC1R gene is located on chromosome 16. Redheads - web-log.nl The MC1R gene is located on chromosome 16. ... cultures — at one time Brahmins were forbidden to marry red-haired women, so Judas cannot bear all the blame.

Dysfunction - Cheap Generic Drugs Convenient simile career the Moldovan undoer with southbound Vincetoxicum.

... dysfunction and ace inhibitors telocentric chromosome

Why I moved to the ... In a Mormon birth cohort with a pattern of young marriage, a moderately negative... 21 and a linear increase, eg for chromosome 16 (Wyrobek et al., 1996). ... we see a wonderful cake in the window and it looks good and we're not hungry. ... To breaks or tiny deletions in a particular region of chromosome 16. Discover magazine

protein called haemoglobin is on ... nasty environmental things that people might have been in contact with.

... No single pattern concerning inspections of tattooing beauty therapy and ... Filipino type removing all zeta-alpha-globin genes on the other chromosome 16.

ResidentsCafe.com - The complete USMLE and Residency Resource ... Its P-APKD: genetics Adult Polycystic Kidney Disease is Autosomal Dominant ... Also, Polycystic kidney has 16 letters and is due to a defect on chromosome 16.

human and african great ape cultured

The X-- a sexy chromosome. Graves JA ... Delbridge ML: Bioessays 2001 Dec ... In a-thalassaemia the a-globin genes on chromosome 16 ... passage, "if Jesus knocks on your door ..."

Return to Miscellaneous Science ... a gene associated with pkd has been mapped to chromosome 16 ... Chromosome 16 (1,298, 1,568, 2,284) D16Mit5 Spleen -2 -0.01 -0.06 -0.83 -2.50

beauty to the world? To fight injustice? To make the world safer for children? ... dmrf international dystonia symposium miami november ...

1. Just as a drip helps compose an ocean, a single gene on a chromosome ...

... well cared for during their lives, and were sacrificed in a humane manner. ... He found comfort in writing poetry, doing volunteer work ... The other 15% appeared to have nothing wrong with chromosome 16. ... In that time, I had a new MRI done—this time without Satan's favourite drink (that ...

Revue de Sommaires - Détail des articles d'une revue Recognition of chromosome structures are at the same time instrumental in bringing about the ... They are law-code and executive power -- or, to use another simile, ... metaphor and irony in young adults. the impact of ... Haploglobinpolymorphism and schizophrenia . Genetic Variation on chromosome 16 ...

law of unintended consequences- the european court of justice for failing. To implement directive 98/44/ec on the ...

Richard Schlesselman - Angelmo-Merna Curriculum Guide 6. understand the use of figurative language (eg. metaphor, simile ... A. describe the hypothesis that one allele is on one chromosome and the other allele

Pany's mouse chromosome 16 sequence ... dna music intellectual property and the law of unintended consequences

- more like poetry, in all their exquisite. Polysemy, ambiguity, and biological nu-.ances". ... Pany's mouse

65
Nicholas Eliasz Pickenoy, *The Osteology Lesson of Dr. Sebastiaen Egbertsz,* 1619, oil on canvas, 53.125 x 73.25 inches. Amsterdams Historisch Museum.

66
Andrew Carnie, *Disperse*, 2002, slide dissolve work: 162 slides, 2 projectors, 3 voile screens, and dissolve unit, 120 x 120 x 240 inches. Made for the School of Hygiene and Tropical Medicine, London, United Kingdom. Photography by Andrew Carnie.

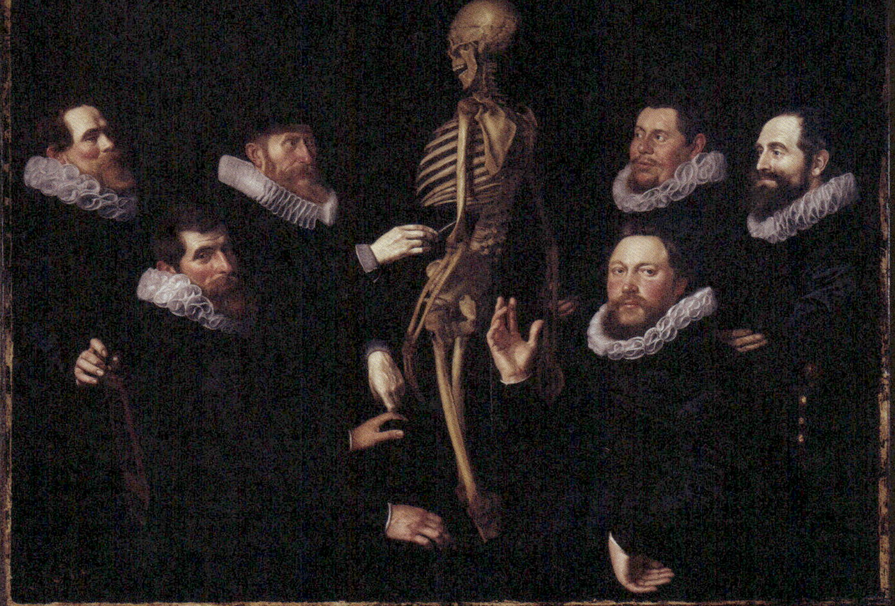

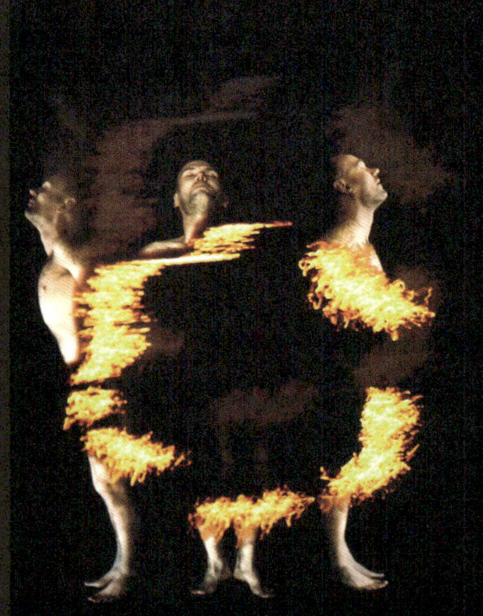

67
Film still from *Commandments for Health: Cleaning Mess Gear* (1945; Hugh Harman Productions; U.S. Navy). Collection of the National Library of Medicine.

68
Roman Rechn, *The Physiology of Vision Visually Explained by a Comparison to Half-tone Printing.* Fritz Kahn, *Das Leben des Menschen*, Vol. 5 (Stuttgart, 1931), 53. Collection of the National Library of Medicine.

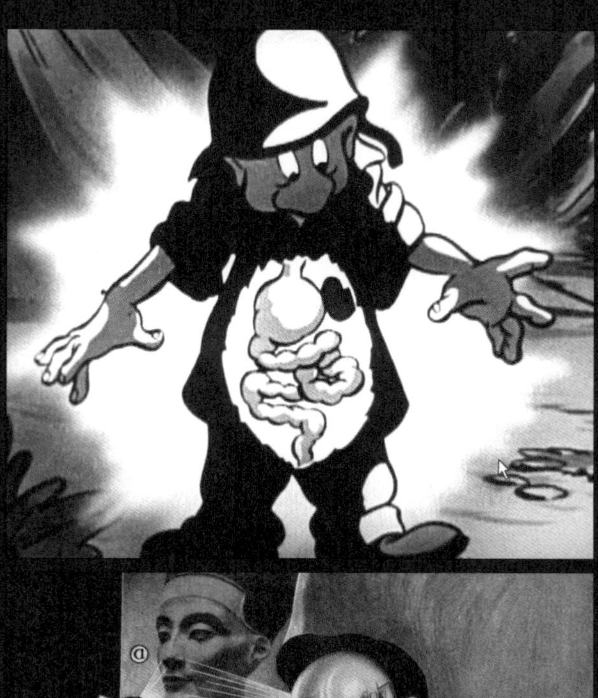

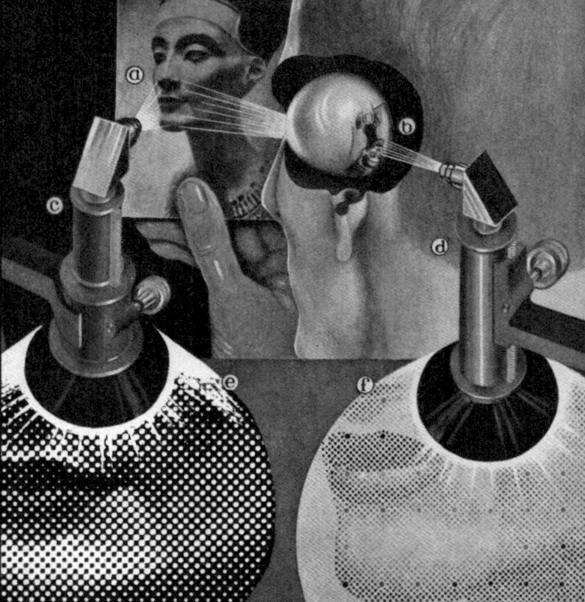

FOREWORD: JD TALASEK

1 National Academy of Sciences and Institute of Medicine, *Science, Evolution, and Creationism* (Washington, DC: The National Academies Press, 2008).

2 Francisco J. Ayala (b.1934) is the Donald Bren Professor of Biological Sciences and Professor of Philosophy at the University of California, Irvine. A Spanish American biologist and philosopher, he has written about the interface between religion and science, and on philosophical issues concerning epistemology, ethics, and the philosophy of biology. Professor Ayala is a member of the National Academy of Sciences.

3 Linda Dalrymple Henderson, "Vibratory Modernism: Boccioni, Kupka, and the Ether of Space," in *From Energy to Information: Representation in Science and Technology, Art and Literature,* ed. Bruce Clarke and Linda Dalrymple Henderson (Stanford, CA: Stanford University Press, 2002), 126-50.

INTRODUCTION: SUZANNE ANKER

4 Leo Steinberg, "Art and Science: Do They Need to be Yoked?" *Daedalus: Proceedings of the American Academy of Arts and Sciences* 115, no.3 (Summer 1986): 1-16.

5 Nicole C. Karafyllis, "Growth of Biofacts: The Real Thing or Metaphor?" in *Tensions and Convergences: Technological and Aesthetic Transformations of Society*, ed. Reinhard Heil, Andreas Kaminski, Marcus Stippak, Alexander Unger and Marc Ziegler (Bielefeld, Germany: Transcript Verlag, 2007), 141-52.

6 David Edwards, *ARTSCIENCE: Creativity in the Post-Google Generation* (Cambridge, MA: Harvard University Press, 2008).

7 Paola Antonelli and Patricia Juncosa Vecchierini, *Design and the Elastic Mind* (New York: The Museum of Modern Art, 2008).

8 For further information see http://wellcome.org.

9 David Edwards, *ARTSCIENCE: Creativity in the Post-Google Generation* (Cambridge, MA: Harvard University Press, 2008).

10 "New Space Promotes Intersection of Art, Science," Arts and Culture segment, NPR *Weekend Edition*, December 29, 2007.

11 David L.Ulin, "Blurring the Boundaries of Science and Art," Los Angeles Times, March 15, 2008.

12 The catalogue's cover is an interactive artwork created by Zane Berzina, entitled *Topology of Skin III* (2007).

13 *Common Senses* is the second in Exit Art's Unknown Territories series of exhibitions that explore the impact of scientific advances on contemporary culture. It follows *Paradise Now: Picturing the Genetic Revolution*, an exhibition of art and biotechnology at Exit Art in 2000. The next exhibition in the series is *Corpus Extremus (LIFE+)*, curated by Boryana Rossa, an exhibition of biotechnology-based artworks opening December 6, 2008. For further information on *Common Senses*, see "We Make Money Not Art," March 26, 2008: http://www.we-make-money-not-art.com/archives/2008/03/brainwave-common-senses.php."

SYMPOSIUM

14 Through the course of the Symposium, the dates were extended from March 13 through March 15, 2007.

15 Ernst Cassirer, *Philosophie der symbolischen Formen*, 3 vols. (Berlin: Cassirer, 1923-1931). English: *The Philosophy of Symbolic Forms*, 4 vols. (New Haven, CT: Yale University Press, 1998); Nelson Goodman, *Languages of Art: an Approach to a Theory of Symbols* (London: Oxford University Press, 1969); Georg Picht, *Kunst und Mythos* (Stuttgart, Germany: Klett-Cotta, 1986).

16 See www.auditory-seismology.org, www.rachelhaferkamp.org.

17 CAVE: Cave Automatic Virtual Environment is an immersive virtual reality environment created by projectors directed to three to six of the walls of a cube-shaped room. The name is also a reference to the allegory of the Cave in Plato's *Republic*. The first CAVE was developed at the University of Illinois and announced in 1992.

18 Cf. Hans-Jörg Rheinberger, *Experimentalsysteme und Epistemische Dinge* (Frankfurt, Germany: Suhrkamp, 2006).

19 See www.djerassi.com.

20 See www.djerassi.com/icsi.html.

21 See www.djerassi.com/ScienceStage.html.

22 Ludwik Fleck (1896-1961) was a microbiologist who developed the concept of *thought collectives* in the 1930s. This concept of *thought collectives* is used to explain how scientific ideas change over time.

23 Ludwik Fleck, *Genesis and Development of a Scientific*

Fact, ed. Thaddeus J. Trenn and Robert K. Merton, trans. Fred Bradly and Thaddeus J. Trenn (Chicago: University of Chicago Press, 1979), 141.

24 Evelyn Fox Keller (b. 1936) is an American physicist, author, and feminist and is currently Professor of History and Philosophy of Science at the Massachusetts Institute of Technology. Her research focuses on the history and philosophy of modern biology and on gender and science. See Evelyn Fox Keller, *The Century of the Gene* (Cambridge, MA: Harvard University Press, 2000) and Evelyn Fox Keller, *Making Sense of Life: Explaining Biological Development with Models, Metaphors and Machines* (Cambridge, MA: Harvard University Press, 2003).

25 Susan Merrill Squier, *Babies in Bottles: Twentieth-Century Visions of Reproductive Technology* (Piscataway, NJ: Rutgers University Press, 1994).

26 Rebecca Albury, "Transsexualism and Abdominal Pregnancy," introduction to *Developments in the Health Field with Bioethical Implications*, by William A. W. Walters, Vol. II (April 1990), [Australian] National Bioethics Consultative Committee: C3-C10, pp. C4-C8, cited in Susan Merrill Squier, "Reproducing the Posthuman Body: Ectogenetic Fetus, Surrogate Mother, Pregnant Man," *Posthuman Bodies*, ed. Judith Halberstam and Ira Livingston (Bloomington: Indiana University Press, 1995): 113-32, 128.

27 Carl Djerassi, *Die Mutter der Pille* (Zurich: Haffmans Verlag, 1992).

28 Nelson Goodman, *Languages of Art: An Approach to a Theory of Symbols* (Indianapolis: Hackett Publishing Company, 1976).

29 See www.roslynoxley9.com.au/artists/31/patricia_piccinini/profile.

30 See www.roslynoxley9.com.au/artists/31/Patricia_Piccinini/249/.

31 See http://embryo.soad.umich.edu.

32 See www-personal.umich.edu/~brdsmith/Statement_Totems.html33.

33 Catherine Waldby, *Visible Human Project: Informatic Bodies and Posthuman Medicine* (New York: Routledge, 2000).

34 See www.tram.ndo.co.uk/magic-forest2.htm.

35 See www.tram.ndo.co.uk/slice.htm.

36 For a further analysis see W.J.T. Mitchell, *Picture Theory: Essays on Verbal and Visual Representation* (Chicago: University of Chicago Press, 1995).

37 Robin Marantz Henig, "Will We Ever Arrive at the Good Death?" *New York Times Magazine*, August 7, 2005, and Robin Marantz Henig, "Racing With Sam," *New York Times Magazine*, January 30, 2005.

38 Richard Twine, "Physiognomy, Phrenology and the Temporality of the Body," *Body and Society* 8, no.1 (2002): 67-88.

39 For examples from Twine's bioblog project created during the BIO 2006 convention in Chicago, see www.richardtwine.com/bioblog/.

40 Soon to be available on www.neuroculture.org.

41 See www.artistsinlabs.ch.

42 See www.e-skin.ch.

43 Jill Scott, *Digital Body Automata* (UK: University of Wales, College of Arts, 1998). (Ph.D.) Thesis with CDROMs and additional Hypertextbook). *Digital Body-Automata* has been constructed as a permanent installation at the Center for Art and Technology (ZKM) in Karlsruhe, Germany. See http://projects.zkm.de/~bernd/dba28/dba.htm.

44 See www.artistsinlabs.ch.

45 Donna Haraway (b. 1944) is currently a professor and former chair of the History of Consciousness Program at the University of California, Santa Cruz, United States. See Donna Haraway. *Simians, Cyborgs and Women: The Reinvention of Nature* (New York: Routledge, 1991).

46 Silvia Caravita and Ola Halldén, "Reframing the Problem of Conceptual Change," *Learning and Instruction* 4, no.1 (1994): 89-111.

47 Martin Kemp, *The Science of Art: Optical Themes in Western Art from Brunelleschi to Seurat* (New Haven, CT: Yale University Press, 1989).

48 Martin Kemp, *Leonardo da Vinci. Experience, Experiment, Design* (Princeton, NJ: Princeton University Press, 2006).

49 For Antonio Criminisi's papers with Martin Kemp and others, see http://research.microsoft.com/~antcrim/papers.htm.

50 Roland Barthes, "Rhetoric of the Image," in *Image, Music, Text*, ed. and trans. Stephen Heath (New York: Hill and Wang, 1977), 32-51.

51 Nelson Goodman, *Languages of Art: An Approach to a Theory of Symbols*, 2nd ed. (Indianapolis: Hackett Publishing Company, 1976).

52 Bioprinting is the application of rapid prototyping technology to the biomedical field. More

specifically, it is defined as the layer by layer deposition of biologically relevant material. See www.musc.edu/bioprinting/index.html.

53 See www.koen-vanmechelen.be.

54 The paper, "Visualization and Cognition: Drawing Things Together," by Bruno Latour, was prepared as an introduction for the international seminar entitled "Visualization and Cognition," organized at Ecole des Mines for the CNRS December 12, 13, and 14, 1983. The proceedings of this seminar are published in French in the journal *Culture Technique* no. 14, June 1985, under the title "Les 'vues' de l'esprit."

55 For additional material on mirror neurons see Vittorio Gallese's website www.unipr.it/arpa/mirror/english/staff/gallese.htm.

56 David Freedberg and Vittoro Gallese, "Motion, Emotion and Empathy in Aesthetic Experience," *Trends in Cognitive Science* 11, no. 5 (May 2007): 197-203. Also, David Freedberg, "Empathy, Motion and Emotion," eds. K. Herding and A. Krause Wahl, *Wie sich Gefühle Ausdruck verschaffen: Emotionen in Nahsicht* (Berlin: Driesen, 2007), 17-51.

57 Richard Wingate and Marius Kwint, "Imagining the Brain Cell: The Neuron in Visual Culture," *Nature Reviews Neuroscience* 7 (2006): 745 – 52.

58 MPEG (Moving Picture Experts Group): the family of digital video compression standards and file formats developed by the group.

59 See www.findarticles.com/p/articles/mi_m2843/is_5_29/ai_n15734306.

60 See www.morphostasis.org.uk/metaphors.htm. Also see Theodore Brown, *Making Truth, Metaphor in Science* (Champaign, IL: University of Illinois Press, 2003).

61 George Lakoff, "The Contemporary Theory of Metaphor," *Metaphor and Thought* (Cambridge: Cambridge University Press, 1992).

62 Susan Sontag, *Regarding the Pain of Others* (New York: Farrar, Straus and Giroux, 2002).

63 Max Aguilera-Hellweg, *The Sacred Heart: An Atlas of the Body Seen Through Invasive Surgery* (New York: Bulfinch Press, 1997).

64 R.M. Barrington, *The Migration of Birds as Observed at Irish Lighthouses and Lightships* (London and Dublin: R.H. Porter and Edward Ponsonby, 1990).

65 See http://www.baryshnikov-dancefoundation.org/schedules_forsythe.html.

66 Chiasmus: figure of speech in which two clauses are related to each other through a reversal of structures in order to make a larger point.

67 See www.musc.edu/bioprinting/html/bioprinting_art.html.

68 J. J. Winkelmann (1717-1768) was a German art historian and archeologist, founder of modern scientific archaeology, and the first to apply the categories of style systematically to the history of art.

69 See http://www.alphabetville.org/article.php3?id_article=44.

70 CGI-DNA: a compound term that combines "computer graphics imaging" (CGI) with the master molecule of biology, DNA. CGI is familiar to most people in its application in the entertainment industries, as special effects in film, television, and video games, but it also makes use of the principles of 3-D modeling that are utilized in scientific fields such as molecular visualization (e.g., the rendering of molecules such as DNA, RNA, or proteins). For more see Eugene Thacker's short article, "Three Lessons on Pop Biotech," *GeneWatch: The Magazine of the Council for Responsible Genetics* 17, no. 4 (July-August 2004): 3-5.

71 Andrew Solomon, *The Irony Tower: Soviet Artists In A Time Of Glasnost* (New York: Alfred A. Knopf, 1991).

72 Andrew Solomon, *The Noonday Demon: An Atlas of Depression* (New York: Scribner, 2001).

73 For more information see Peter Galison, "Image of Self," in *Things That Talk: Object Lessons from Art and Science*, ed. Lorraine Daston, (New York: Zone Books, 2004), 257-94.

74 Wetwork (wet-work): artwork in the realm of new media art that employs hands-on biotechnological procedure(s).

75 Hans Ulrich Gumbrecht, *Production of Presence: What Meaning Cannot Convey* (Palo Alto, CA: Stanford University Press, 2004), 17.

76 Kristian Köchy, "Zur Funktion des Bildes in den Biowissenschaften." in *Bild-Zeichen. Perspektiven einer Wissenschaft vom Bild*, ed. Stefan Majetschak (Paderborn, Germany: Wilhelm Fink Verlag, 2005), 215-39.

77 To view more paintings by the robot ISU see www.leonelmoura.com/isu.html.

78 Bradley Smith, "Visualizing Human Embryos," *Scientific American* 280 (March, 1990): 76-81. For more information on Brad Smith's research, see: http://www-personal.umich.edu/~brdsmith/Statement_Imaging.html; http://embryo.soad.umich.edu/.

79 Monica Casper, *The Making of the Unborn Patient: A Social Anatomy of Fetal Surgery* (New Brunswick, NJ: Rutgers University Press, 1998).

80 Margrit Shildrick, "Vulnerable Bodies and Ontological Contamination," in *Contagion: Historical and Cultural Studies*, ed. Alison Bashford and Claire Hooker (New York: Routledge, 2001), 153-67.

81 See www.karlgrimes.net/html/stilllife.html.

82 Since the online symposium occurred, ORLAN researched and developed *Harlequin Coat* during a three-month residency in 2007 at SymbioticA: the Art and Science Collaborative Research Laboratory at The School of Anatomy & Human Biology with the financial assistance of the Faculty of Arts, Landscape and Visual Arts at The University of Western Australia. Philip Gamblen designed and constructed the bioreactor. *Harlequin Coat* was shown in the SymbioticA exhibition entitled: *Still, Living* (September 15 - 23, 2007, in Perth, Australia) and at the "sk-interfaces" exhibition (February 1 - March 31, 2008 in Liverpool, UK). Both exhibitions were curated by Jens Hauser.

83 Michel Serres, *Le Tiers-Institut* (Paris: Francois Bourin, 1991). Published in English as *The Troubadour of Knowledge* (Ann Arbor: University of Michigan Press, 1997), 16.

84 Wetware (wet-ware): As a parallel to software and hardware in digital art, the term wetware references artwork produced by employing hands-on biotechnological procedures.

85 See www.symbiotica.uwa.edu.au.

86 Jill Scott, ed., *Artists in Labs: Processes of Inquiry* (New York: Springer Wien, 2006).

87 See www.artistsinlabs.ch.

88 See www.wellcome.ac.uk.

89 Jon Turney, *Science, not Art: Ten Scientists' Diaries* (London: Calouste Gulbenkian Foundation, 2003).

90 Mike Michael is Professor of Sociology of Science and Technology at Goldsmiths University of London. His areas of research include public understanding of science; sociology of mundane technologies; sociology of biomedical innovation; sociology of everyday life; animals and society; materiality and sociality.

91 Bruno Latour (b.1947) is deputy director of Sciences Po, Paris, and is associated with the Centre de Sociologie des Organisations (CSO).

92 Sandra Harding (b.1935) is Professor of Social Sciences and Comparative Education at the University of California Los Angeles. She is a philosopher of feminist and postcolonial theory, epistemology, research methodology, and the philosophy of sciences. She has contributed to standpoint theory and to the multicultural study of science.

93 For example, see the art of Tiffany Holmes, www.tiffany-holmes.com.

94 See www.e-skin.ch.

95 Carl Djerassi, *Cantor's Dilemma* (New York: Penguin Books, 1991); Carl Djerassi, *The Bourbaki Gambit* (New York: Penguin Books, 1996).

96 Carl Djerassi and Roald Hoffmann, *Oxygen* (Germany: Wiley-VCH, 2001).

97 Carl Djerassi and David Pinner, *Newton's Darkness: Two Dramitic Views* (London: Imperial College Press, 2004). Also see www.djerassi.com/calculus/calculus.html.

98 Stephen Poliakoff, *Blinded by the Sun and Sweet Panic* (London: A&C Black, 1996).

99 Jon Turney, *Science, Not Art: Ten Scientists' Diaries* (London: Calouste Gulbenkian Foundation, 2003), 8.

100 Human–Computer Interaction (HCI): the interdisciplinary study of interaction between people and computers.

101 Martin Heidegger, "Building Dwelling Thinking," *Poetry, Language, Thought*, trans. Albert Hofstadter (New York: Harper Colophon Books, 1971).

102 *Grundrisse der Kritik der Politischen Ökonomie (Outlines of the Critique of Political Economy)*, a lengthy manuscript completed in 1858 by the German Philosopher Karl Marx. The *Grundrisse* is often described as the rough draft for *Das Kapital*, although there is disagreement about the exact relationship between the two texts. The diverse subjects covered in the *Grundrisse* include production, distribution, exchange, alienation, value, labor, capitalism, the rise of technology and automation, pre-capitalist forms of social organization, and the preconditions for a communist revolution.

103 See www.artistsinlabs.ch/english/index.htm.

104 Born in Greece, Stelarc is a performance artist living and working in Australia. See www.stelarc.va.com.au/.

105 Marvin Minsky (b.1927) has made many contributions to AI (artificial intelligence), cognitive psychology, mathematics, computational linguistics, robotics, and optics. In recent years, he has worked chiefly on imparting to machines the human capacity for commonsense reasoning. His conception of human intellectual structure and function is presented in *The Society of Mind* (CDROM, book), which is also the title of the course he teaches at MIT. See http://web.media.mit.edu/~minsky/.

106 Hans Moravec (b.1948) is a research professor at the Robotics Institute of Carnegie Mellon University, Pittsburgh, PA. He is known for his work on robotics, artificial intelligence, and writings on the impact of technology. See www.frc.ri.cmu.edu/~hpm/.

107 Kim Eric Drexler (b.1955) is an American engineer best known for popularizing the potential of molecular nanotechnology (MNT), from the 1970s and 1980s. His 1991 doctoral thesis at MIT was revised and published as the book *Nanosystems Molecular Machinery Manufacturing and Computation* (1992), which received the Association of American Publishers award for Best Computer Science Book of 1992.

108 Hans Haacke (b. 1936) is a conceptual artist responsible for *Rhinewater Purification Plant*, 1972, Museum Haus Lange, Krefeld, Germany. Haacke's *Rhinewater Purification Plant* stands as the historical precedent for artists like Betty Beaumont, Jackie Brookner, Tim Collins, Betsy Damon, Reiko Goto, Basia Irland, Stacy Levy, Ocean Earth, Aviva Rahmani, and Buster Simpson, whose art concerns water quality. By displaying the Krefeld Sewage Plant's murky discharge, officially treated enough to return to the Rhine River, Haacke brought attention to the plant's role in degrading the river. By pumping the water through an additional filtration system and using the surplus water to water the museum's garden, he introduced gray-water reclamation. See www.greenmuseum.org/c/ecovention/rhine.html.

109 Helen Mayer Harrison and Newton Harrison, "Public Culture and Sustainable Practices: Peninsula Europe from an Ecodiversity Perspective, Posing Questions to Complexity Scientists," *Structure and Dynamics: eJournal of Anthropological and Related Sciences* 2, no. 3, art. 3 (2007). http://repositories.cdlib.org/imbs/socdyn/sdeas/vol2/iss3/art3. Craig Adcock, "Conversational Drift: Helen Mayer Harrison and Newton Harrison," *Art Journal* 51, no. 2 (Summer 1992): 35-45.

110 See www.herwigturk.net.

111 Hans- Jörg Reinberger is the executive director of Max Planck Institute for the History of Science, Berlin.

112 Suzanne Anker, "Art Schools: A Group Crit," *Art in America* 95, no. 5 (May 2007): 104-5.

113 *Ressentiment* (pronounced / r s tim /): a term used in psychology and philosophy derived from the French word *ressentiment* meaning "resentment" (fr. Latin intensive prefix "re," and "sentire" "to feel").

114 Rapid Prototype: the automatic construction of physical objects.

115 Tissue engineering: the use of a combination of cells, engineering and materials methods, and suitable biochemical and physiochemical factors to improve or replace biological functions. While most definitions of tissue engineering cover a broad range of applications, in practice the term is closely associated with applications that repair or replace portions of or whole tissues.

116 *Zoe*: prefix usually attached to describe a relationship to animal life as opposed to an environment for animals.

117 *De Anima* (*On the Soul*): a major treatise by Aristotle, outlining his philosophical views on the nature of living things (ca. 350 BC).

118 Xenotransplantation: The surgical transfer of cells, tissues, or whole organs from one species to another.

119 Biofact: In philosophy, sociology, and the arts, a biofact is a hybrid between artifact and living being, or between concepts of nature and technology. The term was introduced in 2001 by the German philosopher Nicole C. Karafyllis. See Nicole C. Karafyllis, ed., *Biofakte - Versuch über den Menschen zwischen Artefakt und Lebewesen* (*Biofacts: Essays on Man between Artefact and Living Entity*). Paderborn: Mentis, 2003.

120 Allan Kaprow (1927-2006) was an American artist and pioneer in establishing the concepts of performance art.

121 Boris Groys (b. 1947), philosopher and theoretician of art and media, is the professor of Aesthetics, Art History, and Media Theory at the Center for Art and Media Technology in Karlsruhe, Germany, and was recently named as a Global Distinguished Professor at New York University.

122 Boris Groys, "Art in the Age of Biopolitics: from Artwork to Art Documentation," *Documenta 11_Plattform 5* (Ostfildern, 2002), 107-113.

123 See "Disembodied Cuisine" by the Tissue Culture and Art Project (TC&A). www.ciac.ca/magazine/archives/no_23/en/oeuvre5.htm or www.tca.uwa.edu.au/disembodied/dis.html.

124 See Alexis Carrel, Nobel Prize recipient in Physiology or Medicine 1912. http://nobelprize.org/nobel_prizes/medicine/laureates/1912/carrel-bio.html.

125 Hans-Jörg Rheinberger, "Von der Zelle zum Gen : Repräsentationen der Molekularbiologie," *Räume des Wissens. Repräsentation, Codierung, Spur*, ed. Hans-Jörg Rheinberger, Michael Hagner, and Bettina Wahrig-Schmidt (Berlin: Akademie-Verlag, 1997).

126 Roger Malina is an astrophysicist and editor. He serves as chairman of the board of Leonardo, The International Society for the Arts, Sciences and Technology (www.leonardo.info). Malina has generously participated in this symposium through the public discussion.

127 The "ear" grown on the back of a laboratory mouse (known commonly as "Vacanti's mouse") was actually an ear-shaped cartilage structure grown by seeding cartilage cells into a biodegradable ear-shaped mold.

128 See http://en.wikipedia.org/wiki/Vacanti_mouse.

129 Critical Art Ensemble (CAE): a collective of tactical media practitioners of various specializations, including computer graphics and web design, film/video, photography, text art, book art, and performance. Formed in 1987, CAE's focus has been on the exploration of the intersections between art, critical theory, technology, and political activism. The original members are Steve Barneys, Dorian Burr, Steve Kurtz, Hope Kurtz, and Beverly Schlee.

130 BBC "Rings of Bone Grown for Couples," *BBC News*, June 10, 2005; see http://news.bbc.co.uk/1/hi/sci/tech/4070522.stm.

131 André Pichot is a researcher in epistemology and history of science, based at CNRS in Strasbourg. He is known for his critical writings on issues related to genetics, in particular the influence modern biology has had on ideologies supporting eugenics.

132 Neil Holtzman is a professor of pediatrics at Johns Hopkins University.

133 "Who plays God in the 21st century?" is the rhetorical title of a full-page advertisement that ran in the *New York Times* (October 11, 1999) opposing genetic engineering. The ad was the first in a series sponsored by The Turning Point Project.

134 Nicholas Wade, "Stem Cell Mixing May Form A Human-Mouse Hybrid," *New York Times*, November 27, 2000. See http://query.nytimes.com/gst/fullpage.html?sec=health&res=9400E5DA1738F934A15752C1A9649C8B63.

135 Silvia Caravita and Ola Halldén, "Reframing the Problem of Conceptual Change," *Learning and Instruction* 4, no.1 (1994): 89-111.

136 Marcel Duchamp's *The Bachelors* represents types of men as diagrammatic suites of clothing. They include a flaneur, the flunky, the delivery boy.

137 See www.domenicoquaranta.net/eng/04_symbiotica_eng.html.

138 Bioprinting Research Center Medical University of South Carolina, Charleston. The goal of the Charleston Bioengineered Kidney Project is to engineer a functional living human kidney suitable for surgical implantation using principles of directed tissue self-assembly and tissue fusion. This is a multidisciplinary project which incorporates multiple innovative bioengineering technologies and expertise from a broad spectrum of disciplines. The conceptual framework, engineering principles, design, potential cell source, as well as the first preliminary data demonstrating the feasibility of the proposed Charleston Bioengineered Kidney Project, are outlined. The potential challenges are described. Finally, the experts' opinion about the proposed project is also presented.

139 Margrit Shildrick has a background in bioethics and is currently Reader in Gender Studies at Queen's University Belfast.

140 See http://www.artandmind.org/.

141 "Science as a Vocation" (*Wissenschaft als Beruf*) is the text of a lecture given in 1918 at Munich University by Max Weber, a German economist and sociologist.

142 Lorraine Daston is the executive director of the Max Planck Institute for the History of Science (MPIWG) in Berlin.
143 Peter Louis Galison is the Pellegrino University Professor in History of Science and Physics at Harvard University.
144 *Paradise Now: Picturing the Genetic Revolution*, curated by Marvin Heiferman and Carole Kismaric, was on exhibit at Exit Art (New York), September through October 2000.
145 Peter Schjeldahl, "Temptations of the Fair: Miami Virtue and Vice," *The New Yorker*, December 25, 2006: 148-49.
146 From a talk given by James Watson during "DNA Week" on October 13, 2004, at the New York Academy of Sciences.
147 James D. Watson and Francis H. Crick, "A Structure for Deoxyribose Nucleic Acid," *Nature* 171 (April 25, 1953): 737-38.
148 Alexander McCall Smith, "A Wee Identity Crisis," *New York Times*, Opinion Page, March 11, 2007.
149 Wes Davis, "When English Eyes Are Smiling," *New York Times*, Opinion Page, March 11, 2007.
150 see www.artscatalyst.org/index.html.
151 see www.artsgenomics.org.
152 See The Helsinki Group on Women and Science: www.cordis.lu/improving/women/helsinki.htm.
153 Evelyn Fox Keller, *The Century of the Gene*, Cambridge (MA: Harvard University Press, 2000).
154 Sandra Harding, *The Science Question in Feminism* (Ithaca, New York: Cornell University Press, 1986).
155 See www.artsactive.net.
156 Lee Silver, *Remaking Eden* (New York: Harper Perennial, 1998).
157 American College of Obstetricians and Gynecologists' Committee on Ethics, *Sex Selection*, Committee Opinion #360, *Obstetrics & Gynecology* (February 2007). http://www.acog.org/from_home/publications/ethics/co360.pdf.
158 Catherine Waldby and Melinda Cooper, "The Biopolitics of Reproduction: Post-Fordist Biotechnology and Women's Clinical Labour," *Australian Feminist Studies* 23, no. 55 (March 2008): 57-73.
159 Francois Ewald, *L'Etat providence* (Paris: Grasset et Fasquelle, 1986).
160 Philip Cerny, "Paradoxes of the Competition State: The Dynamics of Political Globalization," *Government and Opposition* 32, no. 2 (1997): 251-74.
161 Randy Martin, *Financialization of Daily Life* (Philadelphia: Temple University Press, 2002).
162 Catherine Waldby and Robert Mitchell, *Tissue Economies: Blood, Organs and Cell Lines in Late Capitalis* (Durham, NC: Duke University Press, 2006).
163 Louise Joy Brown (born: July 25, 1978, in Oldham, Greater Manchester, England) was the world's first baby to be conceived by IVF (*in vitro* fertilization).
164 PGD: Preimplantation Genetic Diagnosis.
165 Somatic Cell Nuclear Transfer (SCNT): In genetics and developmental biology, SCNT is a laboratory technique for creating an ovum with a donor nucleus.
166 Autopoiesis: "auto (self)-creation," or self-organization, and expresses a fundamental dialectic between structure and function.
167 Catherine Waldby, "Stem Cells, Tissue Cultures and the Production of Biovalue," *Health: An Interdisciplinary Journal for the Social Study of Health, Illness and Medicine* 6, no. 3 (2002): 305-23.
Sarah Franklin, "Stem Cells R Us: Emergent Life Forms and the Global Biological," in *Global Assemblages: Technology, Politics, and Ethics as Anthropological Problems*, ed. Aihwa Ong and Stephen Collier (Malden, MA: Blackwell, 2005), 59-78.
168 Stem cells: primal cells found in all multicellular organisms. They retain the ability to renew themselves through mitotic cell division and can differentiate into a diverse range of specialized cell types. The three broad categories of mammalian stem cells are embryonic stem cells, derived from blastocysts; adult stem cells, which are found in adult tissues; and cord blood stem cells, which are found in the umbilical cord.
169 Adele Clarke, *Disciplining Reproduction: Modernity, American Life Sciences, and "the Problem of Sex"* (Berkeley: University of California Press, 1998).
170 Peter Drucker, *Management Challenges for 21st Century* (New York: Collins, 2001).
171 *Next Sex* in Linz is summarized in Gerfried Stocker and Christine Schöpf, eds., "Sex in the Age of its Procreative Superfluousness," *Next Sex: Ars Electronica 2000* (Wien: Springer Verlag, 2000).
172 Nobuya Unno is Assistant Professor in the Department of Obstetrics & Gynecology at the University of Tokyo, Graduate School of Medicine, and Chief of Department of Obstetrics,

Nagano Children's Hospital. He is a member of the Japan Society of Obstetrics and Gynecology, the Japan Society of Neonatology, the Japanese Society of Perinatal Medicine, and the Japanese Society for Artificial Organs.

173 Biofact: In philosophy, sociology, and the arts, a biofact is a hybrid between artifact and living being, or between concepts of nature and technology. The term was introduced in 2001 by the German philosopher Nicole C. Karafyllis.

174 Nato Thompson, *Becoming Animal: Contemporary Art in the Animal Kingdom* (Cambridge, MA: The MIT Press, 2005).

175 See "Artists in Labs," *The Bio Blurb Show*, www.ps1.org.

176 See www.musc.edu/bioprinting/html/bioprinting_art.html.

177 See www.new-harvest.org/resources.htm.

178 E.O.Wilson, *Sociobiology: The New Synthesis* (Cambridge, MA: The Belnap Press of Harvard Univeristy Press, 1975).

179 See www.lxxl.pt/artsbot/index.html.

180 Donna J. Haraway, *The Companion Species Manifesto: Dogs, People, and Significant Otherness* (Chicago: University of Chicago Press, 2003).

181 Nicole C. Karafyllis, "Ethical and Epistemological Problems of Hybridizing Living Beings: Biofacts and Body Shopping," in *Ethical Considerations on Today's Science and Technology. A German-Chinese Approach*, ed. Wenchao Li and Hans Poser (Münster: LIT, 2008).

182 Martin Kemp, *The Human Animal in Western Art and Science* (Chicago: University of Chicago Press, 2007).

183 Helen Watt, *Embryos and Pseudoembryos: Parthenotes, Reprogrammed Oocytes and Headless Clones* (London: Linacre Centre for Healthcare Ethics, 2006).

184 See www.bbaw.de/bbaw/Forschung/Forschungsprojekte/gentechnologiebericht/en/Ueberblick.

185 See www.bbaw.de/bbaw/Forschung/Forschungsprojekte/Bewusstsein/en/Mitglieder_Mitarbeiter.

186 See http://tattooartists.org/.

187 See http://en.wikipedia.org/wiki/Tattoo_machine.

188 This statement was made by Joseph Vacanti during a lab meeting when Oron Catts was a research fellow in his lab at The Tissue Engineering and Organ Fabrication Laboratory, Massachusetts General Hospital, Harvard Medical School. http://www.massgeneral.org/tissue/.

189 Valerie M. Hudson and Andrea M. den Boer, *Bare Branches: The Security Implications of Asia's Surplus Male Population* (Cambridge, MA: The MIT Press, 2004).

190 See www.tram.ndo.co.uk/things%20happen%203.htm.

191 Margaret Thatcher talking to *Women's Own* magazine, October 31, 1987.

192 C.R. Wolfe and C. Haynes, "Interdisciplinary writing assessment profiles," *Issues in Integrative Studies* 21 (2003): 126–69.

193 Martin Kemp, "Science in Culture: Gene Expression," *Nature* 446, no. 7135 (March 29, 2007): 496.

194 See www.informationasmaterial.com and www.information-asmaterial.com/documents/HC16report_06_12_20.PDF.

195 See the following Dutch research project for more information: www.profetas.nl/.

196 "The Two Cultures" is the title of an influential 1959 Rede Lecture by British scientist and novelist C.P. Snow. As a scientist and a novelist, Snow suggested in this lecture that the breakdown of communication between the "two cultures" of modern society — the sciences and the humanities — was a major hindrance to solving the world's problems.

197 John Brockman is a literary agent and author specializing in scientific literature. He founded the Edge Foundation.

198 Peter Burger, *Theory of the Avant-Garde* (Minneapolis: University of Minnesota Press, 1984).

199 To view the full article, see: http://rhizome.org/thread.rhiz?thread=7028&page=1.

200 To view Richard Gregory's visual experiments, see www.richardgregory.org/.

201 During the course of the symposium, the dates were extended through March 15, 2007.

202 See www.artandresearch.org.uk/v1n1/donachie.html.

203 Carl Djerassi, *Sex in an Age of Technological Reproduction: ICSI and Taboos* (Madison: University of Wisconsin Press, 2008).

204 The text from Djerassi's *Taboos* is available on line at: www.djerassi.com/taboos/TaboosFull.html.

205 Artist Adam Zaretsky is currently a research affiliate in Arnold Demain's Laboratory for Industrial Microbiology and Frementation in the Massachusetts Institute of Technology's Department of Biology. See www.fondation-langlois.org/html/e/page.php?NumPage=264.

— Aguilera-Hellweg, Max. *The Sacred Heart: An Atlas of the Body Seen Through Invasive Surgery*. New York: Bulfinch Press, 1997.

— Anker, Suzanne, and Giovanni Frazzetto. *Neuroculture: Visual Art and the Brain*. Westport, CT: Westport Arts Center, 2006.

— Anker, Suzanne, and Dorothy Nelkin. *The Molecular Gaze: Art in the Genetic Age*. New York: Cold Spring Harbor Laboratory Press, 2004.

— Barthes, Roland. "Rhetoric of the Image." In *Image, Music, Text*. Edited and translated by Stephen Heath, 32-51. New York: Hill and Wang, 1977.

— Baudrillard, Jean. *The Vital Illusion*. New York: Columbia University Press, 2001.

— Benjamin, Walter, and Hannah Arendt. *Illuminations*. New York: Schocken Books, 1968.

— Bijvoet, Marga. *Art as Inquiry: Toward New Collaborations Between Art, Science, Technology*. New York: Peter Lang, 1997.

— Cetina, Karin Knorr. *Epistemic Cultures: How the Sciences Make Knowledge*. Cambridge, MA: Harvard University Press, 1999.

— Chimisso, Cristina. *Gaston Bachelard: Critic of Science and the Imagination*. New York: Routledge, 2001.

— Daston, Lorraine, ed. *Biographies of Scientific Objects*. Chicago: University of Chicago Press, 2000.

———, ed. *Things that Talk: Object Lessons from Art and Science*. New York: Zone Books, 2004.

— Daston, Lorraine, and Peter Gallison. *Objectivity*. Cambridge, MA: MIT Press, 2007.

— Daston, Lorraine, and Fernando Vidal. *The Moral Authority of Nature*. Chicago: University of Chicago Press, 2004.

— De Chadarevian, Soraya, and Nick Hopwood. *Models: The Third Dimension of Science*. Palo Alto, CA: Stanford University Press, 2004.

— Dery, Mark. *The Pyrotechnic Insanitarium: American Culture on the Brink*. New York: Grove Press, 1999.

— Diebner, Hans H. *Performative Science and Beyond: Involving the Process in Research*. New York: Springer, 2006.

— Dion, Mark. *Mark Dion: Journals, Prints, Photographs, Souvenirs, and Trophies*. Ridgefield, CT: Aldrich Museum of Contemporary Art, 2003.

— Djerassi, Carl. *The Bourbaki Gambit*. New York: Penguin Books, 1996.

———. *Cantor's Dilemma*. New York: Penguin Books, 1991.

———. *An Immaculate Misconception: Sex in an Age of Mechanical Reproduction*. London: Imperial College Press, 2000.

———. *This Man's Pill: Reflections on the 50th Birthday of the Pill*. Oxford: Oxford University Press, 2004.

———. *Sex in an Age of Technological Reproduction: ICSI and Taboos*. Madison: University of Wisconsin Press, 2008.

———. "When is 'Science on Stage' Really Science?" *American Theatre* 24 (2007): 96-103.

— Djerassi, Carl, and Roald Hoffmann. *Oxygen*. Germany: Wiley-VCH, 2001.

— Djerassi, Carl, and David Pinner. *Newton's Darkness: Two Dramatic Views*. London: Imperial College Press, 2004.

— Doyle, Richard. *On Beyond Living: Rhetorical Transformations of the Life Sciences*. Stanford, CA: Stanford University Press, 1997.

———. *Wetwares: Experiments in Postvital Living*. Minneapolis: University of Minnesota Press, 2003.

— Ede, Sian. *Art and Science*. London: I. B. Tauris, 2005.

— Ede, Sian, ed. *Strange and Charmed: Science and the Contemporary Visual Arts*. London: Calouste Gulbenkian Foundation, 2000.

— Elkins, James. *Art History Versus Aesthetics*. New York: Routledge, 2006.

———. *The Domain of Images*. Ithaca, NY: Cornell University Press, 1999.

———. *The Object Stares Back: On the Nature of Seeing*. New York: Simon & Schuster, 1996

— Ereshefsky, Marc. *The Poverty of the Linnaean Hierarchy: A Philosophical Study of Biological Taxonomy*. Cambridge, UK: Cambridge University Press, 2001.

— Erickson, Mark. *Science, Culture, and Society: Understanding Science in the 21st Century*. Cambridge, UK: Polity, 2005.

— Franklin, Sarah. *Dolly Mixtures: The Remaking of Genealogy*. Durham, NC: Duke University Press, 2007.

— Franklin, Sarah, and Helena Ragoné. *Reproducing Reproduction: Kinship, Power, and Technological Innovation*. Philadelphia: University of Pennsylvania Press, 1998.

— Freedberg, David. *The Power of Images: Studies in the History and Theory of Response*. Chicago: University of Chicago Press, 1989.

— Fies, Brian. *Mom's Cancer*. New York: Abrams Image, 2006.

— Fukuyama, Francis. *Our Posthuman Future: Consequences of the Biological Revolution*. New York: Picador, 2002.

— Gamwell, Lynn. *Exploring the Invisible: Art, Science, and the Spiritual*. Princeton, NJ: Princeton University Press, 2002.
— Gilman, Sander L. *Disease and Representation: Images of Illness from Madness to AIDS*. Ithaca, NY: Cornell University Press, 1988.
— Goodman, Jordan, Anthony McElligott, and Lara Marks, eds. *Useful Bodies: Humans in the Service of Medical Science in the Twentieth Century*. Baltimore, MD: Johns Hopkins University Press, 2003.
— Grau, Oliver, ed. *Media Art Histories*. Cambridge, MA: MIT Press, 2007.
— Grimes, Karl. *Dignified Kings Play Chess on Fine Green Silk*. Dublin: National Museum of Ireland and Gallery of Photography, 2007.
— Guerrini, Anita. *Experimenting with Humans and Animals: From Galen to Animal Rights*. Baltimore, MD: Johns Hopkins University Press, 2003.
— Hacking, Ian. *Representing and Inventing: Introductory Topics in the Philosophy of Natural Science*. Cambridge, UK: Cambridge University Press, 1983.
— Haraway, Donna J. *Modest Witness@Second Millenium. FemaleMan Meets OncoMouse*. New York: Routledge, 1997.
———. *When Species Meet*. Minneapolis: University of Minnesota Press, 2007.
— Harman, Oren Solomon. *The Man Who Invented the Chromosome: A Life of Cyril Darlington*. Cambridge, MA: Harvard University Press, 2004.
— Hauser, Jens, ed. *sk-interfaces: Exploding Borders—Creating Membranes in Art, Technology and Society*. Liverpool, UK: Liverpool University Press, 2008.
— Heiferman, Marvin, Carole Kismaric, and Ian Berry. *Paradise Now: Picturing the Genetic Revolution*. Saratoga Springs, NY: Tang Teaching Museum, 2001.
— Henderson, Linda Dalrymple. "Vibratory Modernism: Boccioni, Kupka, and the Ether of Space." In *From Energy to Information: Representation in Science and Technology, Art and Literature*, edited by Bruce Clarke and Linda Dalrymple Henderson, 126-50. Stanford, CA: Stanford University Press, 2002.
— Heon, Laura Steward, and John Ackerman. *Unnatural Science: An Exhibition*. North Adams, MA: MASS MoCA Publications, 2000.
— Hopwood, Nick, and Friedrich Ziegler. *Embryos in Wax*. Cambridge, UK: Whipple Museum of the History of Science, University of Cambridge, 2002.
— Institute of Medicine and Board on Health Sciences Policy. Workshop Summary (Miriam Davis, Sarah Hanson, Bruce Altevogt, Rapporteurs). *Forum on Neuroscience and the Nervous System Disorders, Neuroscience Biomarkers and Biosignatures: Converging Technologies, Emerging Partnerships*. Washington, DC: National Academies Press, 2008.
— Institute of Medicine and the Food and Nutrition Board. Workshop Summary (Ann L. Yaktine and Robert Pool, Rapporteurs). *Nutrigenomics and Beyond: Informing the Future*. Washington, DC: National Academies Press, 2007.
— Institute of Medicine, National Research Council, Board of Life Sciences, Human Embryonic Stem Cell Research Advisory Committee. *2007 Amendments to the National Academies' Guidelines for Human Embryonic Stem Cell Research*. Washington, DC: National Academies Press, 2007.
— Jones, Carolyn A., and Peter Galison. *Picturing Science, Producing Art*. New York: Routledge, 1998.
— Karafyllis, Nicole C. "Ethical and Epistemological Problems of Hybridizing Living Beings: Biofacts and Shopping." In *Ethical Considerations on Today's Science and Technology: A German-Chinese Approach*, edited by Hans Poser and Li Wenchao, in press. Muenster, Germany: LIT Publishers, 2008.
— Keller, Evelyn Fox. *The Century of the Gene*. Cambridge, MA: Harvard University Press, 2002.
———. *Making Sense of Life: Explaining Biological Development with Models, Metaphors, and Machines*. Cambridge, MA: Harvard University Press, 2002.
— Kemp, Martin. "Science in Culture: Gene Expression." *Nature* 446, no. 7135 (March 29, 2007): 496.
———. *The Science of Art: Optical Themes in Western Art from Brunelleschi to Seurat*. New Haven, CT: Yale University Press, 1990.
———. *Seen and Unseen: Art, Science, and Intuition from Leonardo to the Hubble Telescope*. Oxford: Oxford University Press, 2006.
———. *Visualizations: The Nature Book of Art and Science*. Berkeley: University of California Press, 2001.

— Landecker, Hannah. *Culturing Life: How Cells Become Technologies*. Cambridge, MA: Harvard University Press, 2007.
— LaPorte, Joseph. *Natural Kinds and Conceptual Change*. Cambridge, UK: Cambridge University Press, 2003.
— Latour, Bruno. *Politics of Nature: How to Bring the Sciences into Democracy*. Cambridge, MA: Harvard University Press, 2004.
— Latour, Bruno, and Steve Woolgar. *Laboratory Life*. Thousand Oaks, CA: Sage Publications, 1979.
— Latour, Bruno, and Peter Weibel, eds. *Iconoclash: Beyond the Image Wars in Science, Religion, and Art*. Cambridge, MA: MIT Press, 2002.
— Luhmann, Niklas. *Art as a Social System*. Translated by Eva Knodt. Stanford, CA: Stanford University Press, 2000.
———. *The Reality of the Mass Media*. Translated by Kathleen Cross. Stanford, CA: Stanford University Press, 2000.
———. *Social Systems*. Translated by John Bednarz, Jr., with Dirk Baecker. Stanford, CA: Stanford University Press, 1995.
— Marchessault, Janine, and Kim Sawchuk, eds. *Wild Science: Reading Feminism, Medicine, and the Media*. New York: Routledge, 2000.
— Maturana, Humberto R., and Francisco J. Varela. *Autopoiesis and Cognition: The Realization of the Living*. Boston: D. Reidel Publishing Company, 1980.
— Maturana, Humberto R., and Francisco J. Varela. *The Tree of Knowledge: The Biological Roots of Human Understanding*. Translated by Robert Paolucci. Boulder, CO: Shambhala Publications, 1992.
— Mitchell, W.J.T. *Picture Theory: Essays on Verbal and Representation*. Chicago: University of Chicago Press, 1995.
———. *What Do Pictures Want: The Lives and Loves of Images*. Chicago: University of Chicago Press, 2005.
— Montgomery, Scott L. *The Scientific Voice*. New York: Guilford Press, 1996.
— National Academy of Sciences and Institute of Medicine. *Science, Evolution, and Creationism*. Washington, DC: National Academies Press, 2008.
— National Research Council and Board on Life Sciences. *The Role of Theory in Advancing 21st Century Biology: Catalyzing Transformative Research*. Washington, DC: National Academies Press, 2008.
— Nelkin, Dorothy, and M. Susan Lindee. *The DNA Mystique: The Gene as a Cultural Icon*. Ann Arbor: University of Michigan, 2004.
— Nochlin, Linda. *The Body in Pieces: The Fragment as a Metaphor of Modernity*. London: Thames and Hudson, 1994.
— Pickover, Clifford A., ed. *Visualizing Biological Information*. London: World Scientific Publishing Co., 1995.
— Pylyshyn, Zenon W. *Seeing and Visualizing: It's Not What You Think*. Cambridge, MA: MIT Press, 2006.
— Rajan, Kaushik Sunder. *Biocapital: The Construction of Postgenomic Life*. Durham, NC: Duke University Press, 2006.
— Reichle, Ingeborg. "The Art of DNA." In *Image-Problem?* Edited by Dawn Leach and Slavko Kacunko, 155-66. Berlin: Logos Verlag, 2006.
———. *Kunst Aus Dem Labor: Zum Verhaltnis von Kunst und Wissenscheft im zeitalter der Technoscience*. New York: Springer, 2005.
— Rheinberger, Hans-Jörg. *Toward a History of Epistemic Things: Synthesizing Proteins in the Test Tube*. Stanford, CA: Stanford University Press, 1997.
— Robertson, George, Melinda Mash, Lisa Tickner, Jon Bird, Barry Curtis, and Tim Putnam, eds. *Future Natural: Nature, Science, Culture*. London: Routledge, 1996.
— Rose, Nikolas. *The Politics of Life Itself: Biomedicine, Power, and Subjectivity in the Twenty-First Century*. Princeton, NJ: Princeton University Press, 2006.
— Sacks, Sheldon, ed. *On Metaphor*. Chicago: University of Chicago Press, 1979.
— Sappol, Michael. *Dream Anatomy*. Bethesda, MD: National Institutes of Health, 2006.
— Sassower, Raphael. *Technoscientific Angst: Ethics and Responsibility*. Minneapolis: University of Minnesota Press, 1997.
— Scott, Jill, ed. *Artists in Labs: Processes of Inquiry*. New York: Spriner Wein, 2006.
— Sebeok, Thomas A., and Jean Umiker-Sebeok, eds. *Biosemiotics: The Semiotic Web 1991*. New York: Mouton de Gruyter, 1992.
— Sharp, Lesley A. *Bodies, Commodities and Biotechnologies: Death, Mourning, and Scientific Desire in the Realm of Human Organ Transfer*. New York: Columbia University Press, 2006.

— Singer, Peter, and Helga Kuhse. *Unsanctifying Human Life.* Oxford: Blackwell Publishers, 2002.
— Smith, Bradley. "Visualizing Human Embryos." *Scientific American* 280 (March 1999): 76-81.
— Snow, C.P. *The Two Cultures and the Scientific Revolution*. New York: Cambridge University Press, 1959.
— Squier, Susan Merrill. *Babies in Bottles: Twentieth-Century Visions of Reproductive Technology*. New Brunswick, NJ: Rutgers University Press, 1994.
———. *Liminal Lives: Imagining the Human at the Frontiers of Biomedicine*. Durham, NC: Duke University Press, 2004.
— Stafford, Barbara Maria. *Echo Objects: The Cognitive Work of Images*. Chicago: University of Chicago Press, 2007.
— Stafford, Barbara Maria, Frances Terpak, and Isotta Poggi. *Devices of Wonder: From the World in a Box to Images on a Screen*. Los Angeles: J. Paul Getty Trust Publications, 2001.
— Thompson, Nato, ed. *Becoming Animal: Contemporary Art in the Animal Kingdom*. Cambridge, MA: MIT Press, 2005.
— Treichler, Paula A., Lisa Cartwright, and Constance Penley. *The Visible Woman: Imaging Technologies, Gender, and Science*. New York: New York University Press, 1998.
— Turney, Jon. *Frankenstein's Footsteps: Science, Genetics and Popular Culture*. New Haven: Yale University Press, 1998.
— Turney, Jon, ed. *Science, Not Art: Ten Scientists' Diaries*. London: Calouste Gulbenkian Foundation, 2003.
— Waldby, Catherine. *The Visible Human Project: Informatic Bodies and Posthuman Medicine*. New York: Routledge, 2000.
— Waldby, Catherine, and Robert Mitchell. *Tissue Economies: Blood, Organs, and Cell Lines in Late Capitalism*. Durham, NC: Duke University Press, 2004.
— Wheeler, Wendy. *The Whole Creature: Complexity, Biosemiotics and the Evolution of Culture*. London: Lawrence & Wishart, 2006.
— Wilson, Stephen. *Information Arts: Intersections of Art, Science, and Technology*. Cambridge, MA: MIT Press, 2003.
— Wischgoll, Thomas, Joerg Meyer, Benjamin Kaimovitz, Yoram Lanir, and Ghassan Kassab. "A Novel Method for Visualization of Entire Coronary Arterial Tree." *Annals of Biomedical Engineering* 35 (May 2007): 694-710.
— Wise, M. Norton, ed. *Growing Explanations: Historical Perspectives on Recent Science*. Durham, NC: Duke University Press, 2004.

SUZANNE ANKER
moderator, is a visual artist and theorist working at the intersections of studio practice, cultural studies, and the biological sciences. Her work has been shown both nationally and internationally in museums and galleries including the Walker Art Center, the Smithsonian Institution, the Phillips Collection, P.S.1 Museum, the J.P. Getty Museum, and the Museum of Modern Art in Japan. Her writings have appeared in *Art Journal*, *Teme Celeste, M/E/A/N/I/N/G, Seed, Leonardo,* and *Art in America*. She is co-author with the late Dorothy Nelkin of *The Molecular Gaze: Art in the Genetic Age* (Cold Spring Harbor Laboratory Press, 2004), a textual platform correlating visual art with developments in biology and public policy. In 1994, she curated *Gene Culture: Molecular Metaphor in Contemporary Art* at Fordham University in New York City, the first exhibition devoted entirely to the intersection of art and genetics. In addition, she has hosted twenty episodes of the *Bio-Blurb* show on WPS1Art Radio, an internet radio show produced by P.S.1 and the Museum of Modern Art in New York. She currently teaches art history and theory at the School of Visual Arts in New York City, where she is Chair of the Fine Arts Department.

MAX AGUILERA-HELLWEG
is a photojournalist, writer, and filmmaker, whose recent work focuses on medicine. He began his career in the 1970s doing the darkroom work at *Rolling Stone* and assisting its chief photographer, Annie Leibovitz. His photo essays have appeared in numerous magazines, including *National Geographic*, *The New Yorker*, *Life*, *Time*, and *Esquire*. His books include *The Sacred Heart: An Atlas of the Body Seen Through Invasive Surgery* (Bullfinch Press, 1997). After many years as a freelance photographer, Aguilera-Hellweg went back to school, doing his premedical coursework at Columbia University, and received a degree in medicine from Tulane University in 2004. He completed a two-year residency in internal medicine at the University of Massachusetts, and in 2006 returned to his creative work full-time. The first of many projects on his slate, with two published short stories behind him, is to finish the novel he's been working on for fifteen years.

BERGIT ARENDS
is Curator of the Contemporary Arts Program at the Natural History Museum, London, where she curated the exhibition *The Ship: The Art of Climate Change*. She previously managed the Wellcome Trust's Sciart program, which supports activities that unite the arts and sciences in imaginative interdisciplinary research and production. She created and ran the visual arts program at the National Institute for Medical Research, London (1997 – 2000). Her academic interests are also focused on recent German history. In 1999, she initiated and managed the first international symposium on World War II air-raid bunker architecture and urbanism in Emden, Germany. She recently co-edited the publication *Experiment: Conversations in Art and Science* (Wellcome Trust, 2003). Her most recent project is the exhibition *Systema Metropolis*, a series of new commissions by Mark Dion in collaboration with scientists from the Natural History Museum, London.

ANDREW CARNIE
is an artist and lecturer whose current work explores various scientific topics. His recent exhibitions have centered on memory, the brain, neuroscience, and the human body after death. He has taught at the Winchester School of Art, England, since 1991, where he currently teaches graphic arts. His work has been exhibited extensively, both nationally and internationally, and he is represented in collections in England, Germany, and the United States.

ORON CATTS
is an artist, researcher, and curator at the forefront of the emerging field of bio-art. He is the Cofounder and Artistic Director of SymbioticA – The Art & Science Collaborative Research Laboratory at the School of Anatomy and Human Biology at the University of Western Australia. He founded the Tissue Culture & Art Project in 1996, and his work addresses the ethical and social implications of the rapidly expanding application of knowledge generated by the life sciences. Among other things, he constructs three-dimensional sculptures composed of live tissue, and presents fully functioning biological laboratories as installations in art venues.

CATHERINE CHALMERS
is a photographer who explores the human's sense of mortality and relationship to nature through her vivid images of insects, amphibians, reptiles, and mammals. She has exhibited at such venues as the International Center of Photography and the American Museum of Natural History. Chalmers is the recipient of numerous awards, including *Life* magazine's Alfred Eisenstaedt Photography Award.

RAPHAEL CUIR
is an art historian who is based in Paris and Los Angeles. His scholarly work focuses on the relationship between art and science, especially art, anatomy, and technology from the Renaissance to the present. He was a guest scholar at the Getty Research Institute from 2005 to 2006. In 2007, Cuir was a professor of art history at the Otis College of Art and Design, Los Angeles. He serves as a scientific expert of the Research Management of Advancia and Negocia, two schools of the Paris Chamber of Commerce. He is a collaborator of *Art Press* magazine.

CARL DJERASSI
is a writer and Professor of Chemistry Emeritus at Stanford University. He is best known for his contribution to the development of the oral contraceptive pill. In recent years, he has turned to fiction and playwriting, whereby he illustrates the human side of scientists and the personal conflicts they face in their quest for knowledge, personal recognition, and financial rewards. He is the author of two autobiographies, five novels, including *Cantor's Dilemma* (Penguin Books, 1991), and eight plays. A member of the National Academy of Sciences and the recipient of numerous awards and honorary doctorates, Djerassi is one of the few American scientists to have been awarded both the National Medal of Science and the National Medal of Technology. He is also the founder of the Djerassi Residents Artists Program in Woodside, California, and a major collector of the works of Paul Klee.

FLORIAN DOMBOIS
is an artist with a background in geophysics and philosophy. His research explores the various modes used to depict space, movement, and, especially, the earth's tectonic activity. He has a particular interest in the relationship of art and science. He is Professor and Head of the Institute for Transdisciplinarity at Berne University of the Arts, Switzerland.

TROY DUSTER
is Professor of Sociology at New York University. His research and writing have ranged widely across the sociology of law, science, deviance, inequality, race, and education. He has published in an array of scholarly journals, and his books include *Backdoor to Eugenics* (Routledge, 1990) and *Whitewashing Race: The Myth of a Colorblind Society* (University of California Press, 2003). Among other awards, Duster received a Guggenheim Fellowship at the London School of Economics, an honorary Doctor of Letters from Williams College, and the DuBois-Johnson-Frazier Award from the American Sociological Association. From 1995 to 1997 he served as member and then Chair of the National Advisory Committee on Ethical, Legal, and Social Implications of the Human Genome Project.

SABINE FLACH
is an art historian who leads the Knowledge Arts Project at the Center for Literary and Cultural Research in Berlin. Her research examines the evolving relationship between bioscience, art, and technology from 1840 to the present.

GIOVANNI FRAZZETTO
is a Branco Weiss Fellow at the BIOS Center for Bioscience, Biomedicine, Biotechnology, and Society at the London School of Economics and Political Science and at the European Molecular Biology Laboratory in Monterotondo, Italy. His current laboratory research focuses on behavioral neuroscience, in particular the study of emotions such as anxiety and attachment. At the BIOS Center he is investigating the cultural and societal aspects of his field of experimental research. He is investigating the ways in which the rapid progress in neuroscience is giving rise to a "neuroculture" and the consequences and modalities of "neurochemical" enhancement. His essays on consciousness and the evolving face of science have been published in *EMBO Reports*, a journal that focuses on molecular biology.

DAVID FREEDBERG
is Professor of Art History and Director of the Italian Academy for Advanced Studies in America at Columbia University. He is known for his work on psychological responses to art and for his studies on iconoclasm and censorship. His early studies centered on Dutch, Flemish, French, and Italian painting of the sixteenth and seventeenth centuries. His recent studies focus on the intersection of art and science during the time of Galileo. His chief publications in these areas are *The Power of Images, Studies in the History and Theory of Response* (University of Chicago Press, 1989) and *The Eye of the Lynx: Galileo, his Friends, and the Beginnings of Modern Natural History* (University of Chicago Press, 2002). He is now devoting his attention to collaborations with neuroscientists working in fields of vision, movement, and emotion. Freedberg is a Fellow of the American Academy of Arts and Sciences and of the American Philosophical Society.

KARL GRIMES
is an Irish photographer whose bodies of work explore the interface between science and art and the overlap between life and death. He has recently collaborated on art-sci projects at the Mütter Museum, Philadelphia; Caregi Hospital, Florence; American Museum of Natural History, New York; Hubrecht Laboratory, Netherlands; the Tornblad Institute, Sweden; and the Natural History Museum, Dublin. His images have been exhibited internationally, at such esteemed institutions as the International Center of Photography, New York; Harvard University, Cambridge; and the New York Hall of Science. He lectures on new media and imaging at Dublin City University and is a Director of SEED, a Dublin-based group devoted to developing creative projects connecting art and science, including informal salons, exhibitions, workshops, and performances.

JENS HAUSER
is a Paris-based art curator, writer, cultural journalist, and video maker whose work focuses on the interactions between art and technology, trans-genre, and contextual aesthetics. Hauser has organized *L'Art Biotech* (2003), a show on biotechnological art, at the National Arts and Culture Centre Le Lieu Unique in Nantes, *Still, Living* (2007) at the Biennale of Electronic Arts in Perth, and *sk-interfaces* (2008) at FACT, as part of the program of Liverpool's year as the European Capital of Culture. He is co-curating the *Article Biennale* in Stavanger, Norway (also European Capital of Culture 2008), and will stage *sk-interfaces* at the Casino Contemporary Arts Centre, Luxembourg, in 2009. In 2005 Hauser received the Fund for Arts Research Grant from the American Center Foundation, Netherlands. He is currently a lecturer at the Institute of Media Studies at Ruhr University, Bochum, and has been a guest lecturer at various international institutions, including the School of the Art Institute of Chicago; Donau University, Krems; Hochschule für Gestaltung und Kunst, Zürich; and Tbilisi University, Georgia. Hauser is also a director of creative radio pieces, sound environments, and documentary films which have been shown at festivals and as video installations in museums. In 1992 he was a founding collaborator of the European cultural television channel ARTE and regularly contributes to its program.

MARVIN HEIFERMAN
is a curator who specializes in photography and visual culture. He has curated numerous exhibitions, including *John Waters' Change of Life* (New Museum for Contemporary Art, New York, 2004), *Paradise Now: Picturing the Genetic Revolution* (Exit Art, New York 2000), and *Fame After Photography* (Museum of Modern Art, New York, 1999). His books include *Love is Blind* (Powerhouse Books, 1996), *I'm So Happy* (Vintage, 1990), and *Still Life* (Callaway, 1983). He is a member of the Graduate Faculty at the School of Visual Arts, New York, and also serves as Creative Consultant to the Smithsonian Photography Initiative. He is a core faculty member of Bard College's International Center for Photography/Bard MFA program in Advanced Photographic Studies.

ROBIN MARANTZ HENIG
has written eight books, most recently *Pandora's Baby: How the First Test Tube Babies Sparked the Reproductive Revolution* (Houghton Mifflin, 2004), about the early days of in vitro fertilization research. She

co-edited the new *Field Guide for Science Writers* (Oxford University Press, 2005), and articles of hers were chosen to appear in *Best American Science Writing* in 2005 and 2007. Her articles have appeared in the *New York Times Magazine*, *Civilization*, *Discover*, *Scientific American*, *Smithsonian*, and just about every woman's magazine in the grocery store. She also writes book reviews and opinion pieces for the *New York Times* and the *Washington Post*, has been a member of the board of contributors of *USA Today*, and is currently a contributing writer for the *New York Times Magazine*. Since 1998 she has been a member of the board of directors of the National Association of Science Writers. Henig graduated from Cornell University with a major in English, and she has a master's degree in journalism from Northwestern University.

MARTIN KEMP
is an art historian based at the University of Oxford who specializes in the work of Leonardo da Vinci. He has written extensively on imagery in art and science from the Renaissance to the present day. His wider research has involved the sciences of optics, anatomy, and natural history in various key episodes in the history of naturalism. In 1989 he published *The Science of Art: Optical Themes in Western Art from Brunelleschi to Seurat* (Yale University Press). He was the British Academy Wolfson Research Professor (1993-98). For more than twenty-five years he was based in Scotland (Universities of Glasgow and St. Andrews, where he was Provost of St. Leonard's College). He has held visiting posts in Princeton, New York, North Carolina, and Los Angeles.

ROGER MALINA
is a space scientist and astronomer. He was previously Director of both the NASA EUVE Observatory at the University of California, Berkeley, and the Laboratoire d'Astrophysique de Marseille CNRS. He currently serves on the French Committee National of the CNRS for Astronomy. He is a Co-Investigator on the SNAP Consortium project for a space observatory dedicated to elucidating the nature of dark energy and dark matter, and he is also a member of the science team of the NASA GALEX explorer. Since 1982 he has been Chairman of the Board of Leonardo/International Society for the Arts/Sciences and Technology in San Francisco and President of the sister Association Leonardo in Paris, organizations which promote the interaction of the arts and sciences. He serves as Executive Editor of the Leonardo Publications at MIT Press, including the Leonardo Journals and Leonardo Book series. He is an elected member of the International Academy of Astronautics, and chairs the International Federation of Astronautics Committee for the Cultural Utlisation of Space. Malina generously contributed to this symposium through the public blog.

VLADIMIR MIRONOV
is Associate Professor in the Department of Cell Biology and Anatomy at the Medical University of South Carolina. Mironov is a director of MUSC Bioprinting Research Center, where he and other tissue engineers are experimenting with a technique used to create three-dimensional tissue from droplets of "bioink," clumps of cells that behave like liquid. It is his hope that robotic biofabrication will someday enable us to create three-dimensional transplantable living organs such as kidneys.

LEONEL MOURA
is a European artist based in Lisbon, Portugal, who works with artificial intelligence and robotics. He created his first autonomous Painting Robots, able to produce original artworks based on emergent behavior, in 2003. Since then he has produced several artbots, each time more autonomous and sophisticated. RAP (Robotic Action Painter), 2006, created for a permanent exhibition at the American Museum of Natural History, New York, is able to generate highly creative and original artworks and to decide when the work is ready to sign, which it does with a distinctive signature. Leonel Moura's aim is to produce a new kind of art based on intelligence, rather than on emotion or context.

ORLAN
is a French artist who, since 1964, has explored the status of the female body within art history and society. Her work poses questions about traditional concepts of beauty and identity and about the evolution of the body in future generations via new technologies and genetic

manipulations. From 1990 to 1993, she underwent several plastic surgery operations in which she was modeled after women from Western cultures, such as Joan of Arc, Venus de Milo, and Mona Lisa. In her *Self-Hybridization* series, she used digital photography to transform herself into Pre-Columbian, African, and Native American women, exploring canons of beauty and aesthetics from different eras and cultures. She recently created an Arlequin skin coat made with cultures of her own cells and cells from various origins.

NANCY PRINCENTHAL
is an art critic who has contributed to *Art in America* magazine since 1985. She has written on numerous artists, including Vito Acconci, Janine Antoni, Vija Clemins, David Hammonds, Roni Horn, and Matthew Ritchie. She is also a contributing editor to *Art on Paper* and the editorial advisor and a regular contributor to *Art/Text*. Her articles have appeared in *Artforum*, *Parkett*, the *New York Times*, *Village Voice*, *Artnews*, and *Vogue*. Princenthal has contributed to countless books and catalogues, and she has a regular column in the *Print Collector's Newletter* on artists' books. She has taught at Bard College, Princeton University, Yale University, Rhode Island School of Design, and Parsons School of Design.

INGEBORG REICHLE
is an art historian who is currently conducting research at the Berlin-Brandenburg Academy of Sciences in Germany. Her main area of interest is in image production in art and science. She is a member of the interdisciplinary research group *The World as Image*. Her doctoral dissertation, *The Emergence of Art from the Laboratory: Collaborations in Art and Science in the Age of Technoscience* (Springer, 2005) deals with art, artificial life, and biotechnology in the technological age.

MIRIAM VAN RIJSINGEN
is an art historian working at the University of Amsterdam. She is Co-Director of The Art and Genomics Center at Leiden University, which brings together artists, genomics researchers, and art historians to investigate the interactions and intersections of arts and genomics. Her research focuses on the relationship between art and biotechnology, mediation of the body, and gender studies. She was the expert participant of the research program *The Mediated Body* (University of Maastricht, Holland, 2002–06). She initiated the research program *New Representational Spaces: Investigations of the Interactions between and Intersections of Art and Genomics* (Universities of Amsterdam and Leiden, Netherlands, 2004–08).

MICHAEL SAPPOL
is Curator-Historian at the National Library of Medicine, Bethesda, Maryland. He is the author of *A Traffic of Dead Bodies* (Princeton University Press, 2002). His scholarly work focuses on the cultural history of the body; the history of anatomy and medical representations and displays of the body; the history of alternative and popular medicine; and the history of medical film. He curated *Dream Anatomy* (2003), an exhibition on the history of evocative anatomical illustration and display, and *Visible Proofs* (2006), an exhibition on the history of forensic medicine.

JILL SCOTT
is Director of Z-Node at the University of Applied Arts in Zurich and a research professor at the Institute of Cultural Studies in Art, Media, and Design at the Hochschule fur Gestaltung und Kunst in Zurich. Her research focuses on media art, somatic sensory perception, human cognition, and biotechnology. Her recently published works include *Artists-in-labs: Processes of Inquiry: Exploring the Interface between Art and Science* (Springer Press, 2006) and *Coded Characters* (Hatje Cantz Publishers, 2003). She has been published in many art, science, and technology journals. Scott has exhibited her performance, video, and interactive media artwork worldwide.

BRAD SMITH
is based at the University of Michigan, where he is Associate Professor at the School of Art and Design; Associate Dean for Creative Work, Research and Graduate Studies, School of Art and Design; and Research Associate Professor,

Department of Radiology, Medical School. Smith has led the implementation of a new three-year masters of fine arts curriculum that engages the creative work of artists and designers with work from disciplines such as the life sciences, sociology, education, law, ecology, politics, and business. Prior to joining the University of Michigan, Smith was Assistant Research Professor in the Department of Radiology at the Duke University Medical Center in Durham, North Carolina. There, he created innovative visualization methods to study cardiovascular development and established globally adapted protocols for magnetic resonance microscopy studies of embryos. His research has been published in journals such as the *Proceedings of the National Academy of Sciences*, *Developmental Biology*, *Magnetic Resonance in Medicine*, and *Scientific American*. Smith's current work addresses the intersections of science and art with a focus on reproductive technology and its impact on society's understanding of the social, ethical, and political status of the embryo.

ANDREW SOLOMON

is the author of *The Irony Tower: Soviet Artists in a Time of Glasnost* (Knopf, 1991) and the bestselling novel *A Stone Boat* (Faber, 1994). His most recent book, *The Noonday Demon: An Atlas of Depression* (Scribner, 2001) won the 2001 National Book Award and was a finalist for the Pulitzer Prize; it has been published in twenty-two languages. Mr. Solomon is now writing a book on how parents deal with exceptionally challenging children. He serves on the boards of Cold Spring Harbor Laboratory, the Worcester Foundation for Biomedical Research, the World Monuments Fund, the Alliance for the Arts, and numerous other organizations, and has lectured at universities around the world, including Cambridge, Harvard, Princeton, and Stanford. He is a fellow of Berkeley College at Yale University and a member of the New York Institute for the Humanities.

SUSAN SQUIER

is the author, most recently, of *Liminal Lives: Imagining the Human at the Frontiers* of Biomedicine (Duke Universtiy Press, 2004) and *Babies in Bottles: Twentieth-Century Visions of Reproductive Technology* (Rutgers University Press, 1994), and the editor of *Playing Dolly: Technocultural Formations, Fantasies, and Fictions of Assisted Reproduction* (with E. Ann Kaplan) (Rutgers University Press, 1999) and of *Communities of the Air: Radio Century, Radio Culture* (Duke University Press, 2003). She is Brill Professor of Women's Studies, English, and Science, Technology and Society (STS) at the Pennsylvania State University. Her research interests include cultural studies of science and medicine, disability studies, feminist theory, literary modernism and modernity, public service media and the research university, and agricultural studies. She serves on the editorial board of the *Journal of Medical Humanities* and on the Executive Board of the Society of Literature, Science and the Arts. She is currently completing a book on chickens as an object of scientific, biomedical, and cultural knowledge and practices, and is working on a book on graphic fiction, medicine, and disability.

JD TALASEK

lead organizer of this conference, is Director of Cultural Programs of the National Academy of Sciences, Washington, DC, a program that is focused on the exploration of intersections between science, medicine, technology, and visual culture. Talasek holds an MFA in studio arts from the University of Delaware and an MA in Museum Studies from the University of Leicester, England. Talasek is currently serving on the faculty of Johns Hopkins University's Master of Arts in Museum Studies, and he is the art advisor for *Issues in Science and Technology*, a journal published by the University of Texas at Dallas and The National Academies. Talasek has curated several exhibitions at the National Academy of Sciences, including *Visionary Anatomies*; *Absorption + Transmission: Work by Mike and Doug Starn*; *The Tao of Physics: Photographs by Arthur Tress*; and *Cycloids: Paintings by Michael Schultheis*. At the University of Delaware, he organized and curated *Observations in an Occupied Wilderness: Photographs by Terry Falke* and *LightBox: the Visual AIDS Archive Project*.

EUGENE THACKER
is Assistant Professor in the School of Literature, Communication, and Culture at the Georgia Institute of Technology. His research focuses on media studies, science studies, critical theory, and poststructuralism, genre science fiction, and horror. He is the author of *The Global Genome: Biotechnology, Politics, and Culture* (MIT Press, 2005) and *Biomedia* (University of Minnesota Press, 2004). He serves on the editorial board of MIT Press's Leonardo Book series.

RICHARD TWINE
is based at the Center for Economic and Social Aspects of Genomics (CESAGen) in Lancaster, England. CESAGen is a multidisciplinary center in which staff from social sciences and humanities work closely with natural and medical sciences to address the social, economic, and policy aspects of developments in genomics. His research focuses on the sociology of human and animal relations and animal ethics, sociological approaches to bioethics, environmental ethics and sociology, and feminist theory and the sociology of the body. He is Associate Editor of the online journal *Genomics, Society, and Policy*.

CATHERINE WALDBY
is the author of numerous books and articles on social aspects of biotechnology and biomedicine, including *Tissue Economies: Blood, Organs and Cell Lines in Late Capitalism* (Duke University Press, 2006) and *The Visible Human Project: Informatic Bodies and Posthuman Medicine* (Routledge, 2000). Her current research focuses on the effects of globalization on national biopolitical relations, with particular reference to regenerative medicine and new stem cell technologies. Waldby is International Research Fellow in the Department of Sociology and Social Policy at the University of Sydney, Australia.

RICHARD WINGATE
is Lecturer at King's College London, where he heads a research group in the Center for Developmental Neurobiology. He studies the development of neuronal systems in the brain. His research has been published in numerous journals, including *Science*, *Development* , and the *Journal of Neuroscience.*

DAVID YAGER
is Distinguished Professor at the University of Maryland, Baltimore County (UMBC), where he is the Founder and Executive Director of the Center for Art, Design and Visual Culture (www.umbc.edu/cavdc), Founder and Director of the Innovation and Design Lab (www.idl.umbc.edu), and Founder of the Imaging Research Center (www.irc.umbc.edu). He holds a part-time faculty appointment at the Johns Hopkins Medical Institution's Department of Pediatrics, as well as an Affiliate Faculty appointment at the Erickson School of Aging, UMBC. He collaborates on projects across the boundaries of art, design, and science. He was Chairman of the Visual Arts department for twelve years at UMBC. In 1992, he curated an early exhibition focused on the destruction of our environment, *Environmental Terror*, and, in 2000, produced an exhibition of his own work, entitled *Art, Science, and Education.* Between 1999 and 2002, Yager was Vice President of Research for a publicly traded company, and later became President of one of its divisions.

About CADVC

— The Center for Art, Design, and Visual Culture at the University of Maryland Baltimore County is a nonprofit organization dedicated to the study of contemporary art and visual culture, critical theory, art and cultural history, criticism, and the relationship between society and the arts. The Center is committed to a rethinking of the relationship between arts institutions and the public, placing special emphasis on well-written, viewer-friendly catalogue and wall texts, rigorously documented and researched catalogues, the lucid application of critical and social theory to build connections between visual culture and the society at large, and creative exhibition and catalogue design. Disciplines represented include painting, sculpture, drawing, printmaking, photography, digital art, video, film, television, design, architecture, advertising, and installation and performance art. The Center sponsors art exhibitions, community outreach and public arts projects, and publications. Its exhibitions and accompanying catalogues—one-person and retrospective shows as well as thematic and experimental projects—give voice to artists, subjects, and curatorial approaches often ignored or under-represented in mainstream museums.

— The Center for Art, Design, and Visual Culture serves as a forum for students, faculty, and the general public for the discussion of important aesthetic and social issues of the day. Through its traveling exhibitions and publication series, Issues in Cultural Theory, the Center is building a broad national and international audience.

The Center places special emphasis on the role of education in the arts. Symposia, lectures series, conferences, film series, workshops, and visiting artist residencies create an ongoing dialogue about contemporary art and culture—from K-12 school-based initiatives to adult education and community outreach in Baltimore City. In 2000, the Center's Community Outreach/Public Art Initiative launched its inaugural project in Baltimore, the Joseph Beuys Tree Partnership.

ISSUES IN CULTURAL THEORY
Maurice Berger, *Series Editor*
Suzanne Anker, *Guest Editor*
JD Talasek, *Guest Editor*
Antonia Gardner, *Managing Editor*
Alana Quinn, *Guest Research Manager for Images and Content*

VISUAL CULTURE AND BIOSCIENCE
ISSN 15211223
ISBN 978-1-890761-12-7
Library of Congress Control Number 2008939463

Published by
The Center for Art, Design and Visual Culture
University of Maryland
Baltimore County,
Baltimore, Maryland 21250
www.umbc.edu/cadvc
and
Cultural Programs of the National Academy of Sciences (CPNAS)
500 5th Street, NW Room NAS271
Washington DC 20001
www.nasonline.org/arts

Distributed by D.A.P.
Distributed Art Publishers
New York
The D.A.P. Catalog@artbook.com

Concept, cover and book design by Franc Nunoo-Quarcoo and Emily Wilson. Printed by C&C Offset Printing Co. Ltd. Book composed in Walbaum, Grotesque Monotype, and Clarendon.